Crystal Growth
THEORY AND TECHNIQUES

Volume 2

Crystal Growth
THEORY AND TECHNIQUES

Volume 2

Edited by

C. H. L. Goodman
Standard Telecommunication Laboratories Ltd.
Harlow, Essex

Springer Science+Business Media, LLC

Library of Congress Cataloging in Publication Data

Goodman, C H L
Crystal growth.

Includes bibliographical references and index.
1. Crystals–Growth. I. Title.
QD921.G66 548'.5 72-95388
ISBN 978-1-4757-4898-7

ISBN 978-1-4757-4898-7 ISBN 978-1-4757-4896-3 (eBook)
DOI 10.1007/978-1-4757-4896-3

Originally published by Plenum Press, New York in 1978
Softcover reprint of the hardcover 1st edition 1978

Contributors

M. Ettenberg, RCA Laboratories, Princeton, New Jersey

R. Falckenberg, Siemens Central Research Laboratory, Munich, West Germany

C. H. L. Goodman, Standard Telecommunication Laboratories, Ltd., Harlow, Essex, England

J. Nishizawa, Research Institute of Electrical Communication, Tohoku University, Sendai; and Semiconductor Research Institute, Sendai, Japan

G. H. Olsen, RCA Laboratories, Princeton, New Jersey

Preface

The science and art of crystal growing continue to flourish; even with increasing understanding of the science, "feel" and skill continue to play their vital part, as was so clearly evidenced at the recent Boston International Conference on Crystal Growth. The aim of this volume, the same as that of the first, is to try to improve understanding by providing detailed discussions of crystal growth techniques and problems that arise with them. The published paper in the specialized literature is too limited a vehicle, by convention and by editorial pressure on length, to discuss matters in detail, yet it is in the small details born of experience that vital information can often lie concealed. A major aim of this series, therefore, has been to encourage contributors to describe rather fully what has been achieved in their special fields.

The next volume of this series is now well underway and plans for Volume 4 are advanced. If you, the reader, feel that some important aspect of crystal growth is being unjustifiably neglected, perhaps you should consider offering a contribution! And even if you do not wish to do that, please do offer criticism—preferably constructive.

I hope that the present volume will prove as useful and interesting to crystal growers as apparently did the first volume of the series; certainly the warm commendations that that volume has elicited were a great encouragement for the present work.

Finally it is a pleasure to thank Standard Telecommunication Laboratories for its continuing support.

C. H. L. Goodman

Standard Telecommunication Laboratories Ltd.
London Road, Harlow, Essex, England, CM17 9NA

Contents

Chapter 1

Growth Effects in the Heteroepitaxy of III–V Compounds

G. H. OLSEN and M. ETTENBERG

1. INTRODUCTION

The use of thin-film epitaxial deposition has considerably advanced the preparation of III–V compound electronic structures, both in the research laboratory and in the commercial market. The devices may be either homoepitaxial, where the substrate and epitaxial layers are of the same material, e.g., GaP on GaP for visible light-emitting diodes (LEDs), or comprised of heteroepitaxial layers where the substrate and epitaxial layers are of differing III–V materials, e.g., $Al_xGa_{1-x}As$ on GaAs for cw injection lasers. It should be noted that, in actuality, almost all epitaxial layers are heteroepitaxial in that there is a lattice-parameter mismatch between the substrate and epitaxial layer even when they are nominally the same material. This mismatch may arise due to a doping density difference between the substrate and epitaxial layer or even a point defect density difference between the two.

The theory of heteroepitaxial interfaces has received considerable attention over the past few decades,[1,2] and many experimental data are available. However, since most experiments are either influenced or dominated by experimental conditions (e.g., growth temperature, vacuum), theories have been tailored to match the experiments (e.g., 2-D vs. 3-D growth), so that there is often considerable variation among the quantitative predictions. However, most of the theories agree qualitatively on the following generalizations about the accommodation of strain between an epitaxial layer and its substrate:

G. H. OLSEN and M. ETTENBERG · RCA Laboratories, Princeton, New Jersey.

1. Initial deposits of the epitaxial layer will be strained elastically so that its lattice constant equals that of the substrate.
2. After a certain critical thickness (h_c) has been exceeded, it is energetically favorable for part of the strain to be accommodated by plastic deformation (e.g., dislocations).
3. For thicknesses much greater than h_c, it is energetically favorable for all of the strain to be accommodated plastically. The equilibrium spacing(s) of these dislocations can be shown[3] to be

$$S = b/\epsilon \tag{1}$$

where b is the misfit-relieving component of the Burger's vector and ϵ is the misfit strain.

A wealth of experimental evidence has shown these predictions to be qualitatively correct although there is considerable variation among the quantitative predictions. Jesser and Kuhlmann-Wilsdorf[4] and Matthews[5] have extended the theoretical treatments of van der Merwe[1] and have developed expressions for h_c. Although these expressions are rather formidable, they can be approximated as

$$h_c \simeq b/2\epsilon \tag{2}$$

for crystals with similar elastic constants. This approximation agrees with the original expressions to within a factor of 2 for many cases, and since there is often more of a variance between experimental and theoretical values, it should be looked upon as a useful "rule of thumb." However, the original expressions should be used whenever the reliability and consistency of the experimental data demand better precision.

These considerations predict the mechanical state of heteroepitaxial crystals at equilibrium but do not deal with the kinetics (e.g., methods of dislocation introduction and distribution) of crystal growth which often dominate. The purpose of this chapter is to consider the relationships between elastic strain, dislocations, and the crystal growth of heteroepitaxial III–V compounds prepared by vapor-phase epitaxy (VPE) and liquid-phase epitaxy (LPE). Molecular beam epitaxy,[6] which is an interesting new growth technique, is not considered herein. Initially, effects primarily associated with elastic strain only are discussed. Following this, effects associated with misfit dislocations (which of course are induced by the presence of elastic strain) are described. Finally, there is a brief discussion of the optimization of transmission photocathodes, a device structure that is very sensitive to the presence of crystal defects.

The topics discussed in this chapter are limited to III–V compounds. However, a clear and concise review of defects in epitaxial films in general has recently been made by Stowell.[7]

2. THE CRYSTAL STRUCTURE AND GROWTH OF III–V COMPOUNDS

2.1. Crystallography of III–V Compounds

Most III–V compounds (except for the nitrogen compounds) crystallize in the zinc blende structure. This structure may be derived by first superimposing two face-centered-cubic (fcc) structures and then displacing one of them one-quarter of the distance along the cube diagonal. If the atoms of the two fcc structures are identical, then the diamond structure (e.g., Si) results. However, if they are different, the zinc blende structure results. Zinc blende and diamond structures have certain features in common. For instance, dislocations in both are typically of the 60° type, lying along ⟨110⟩ directions on a {111} slip plane with a ⟨011⟩-type Burger's vector. In the case of III–V compounds, the group III atoms all lie on one of the fcc substructures while the group V atoms lie on the other. This gives rise to certain asymmetries of the crystal structure which are not present in fcc or diamond structures.[8] For instance, depending on how it is viewed, a {111} plane in a III–V compound is composed of either all group III or all group V atoms. This causes asymmetric etching effects.[9] The asymmetry among the {111} planes gives rise to asymmetries along ⟨110⟩-type directions. Thus, {001} planes in III–V compounds do not possess a fourfold axis of symmetry (as do diamond and fcc structures), which means that in an (001) plane, the [110] direction is not equivalent to the orthogonal $[1\bar{1}0]$ direction. This, in turn, gives rise to asymmetric mechanical properties such as unidirectional cracking[10] and preferential dislocation introduction along one direction (discussed later).

Table 1

Compound	Lattice parameter (Å)
GaP	5.4510
GaAs	5.6534
AlAs	5.6605
InP	5.8694
InAs	6.0584

The lattice constants of some common binary III–V compounds are listed in Table 1. Ternary alloys of these compounds closely obey Vegard's law in that the lattice parameter is a linear function of composition.

2.2. Vapor-Phase Epitaxy (VPE) Growth Technique

2.2.1. Description of the Method

The VPE reactor[11] used to grow the III–V compounds described in this section is shown in Fig. 1. Details of the chemical processes may be found elsewhere.[12] Any alloy of In, Ga, As, and P may be grown in this system. Mass-flow control valves are used on all input lines of the reactant species to ensure stable flow rates. A typical growth process begins by inserting the substrate into the forechamber (with the entry valve closed) and flushing it in H_2. Meanwhile, flows of HCl are established over the metals and the group V hydride gases are introduced. Doping is accomplished by either heating a bucket of zinc (p-type) or by introducing H_2S gas (n-type). Composition is controlled by adjusting one of the alloying species (e.g., HCl over the In for $In_xGa_{1-x}P$ or PH_3 in the case of $GaAs_{1-x}P_x$). When all flows have been established for at least 15 min and the forechamber is properly flushed, the entry valve is opened and the substrate is placed in the preheat zone for 5 min. During the preheat process, a flow of either AsH_3 or PH_3 is passed over the substrate to prevent decomposition, since most III–V compounds dissociate incongruently at 700°C (i.e., the group V element tends to evaporate preferentially[13]). A dramatic demonstration of this deleterious effect may be seen in Fig. 2, which contains a cross-sectional transmission electron microscope (hereafter referred to as XTEM) micrograph of a n–$In_xGa_{1-x}P$/GaAs/p–$In_xGa_{1-x}P$ double-heterojunction laser.[14] The GaAs substrate was preheated in a pure H_2 atmosphere (without AsH_3)

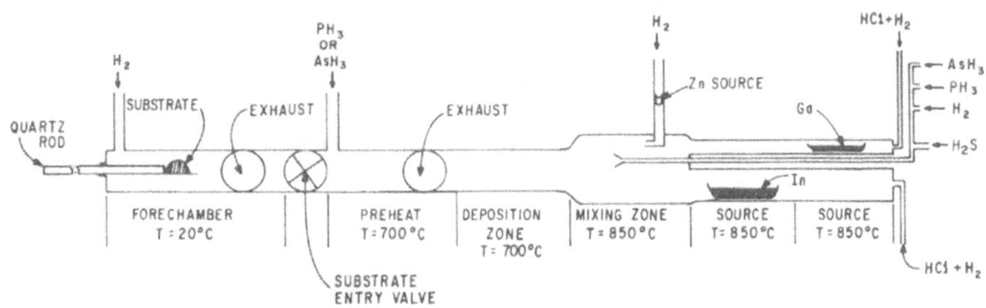

FIGURE 1. RCA vapor-phase epitaxy growth system.

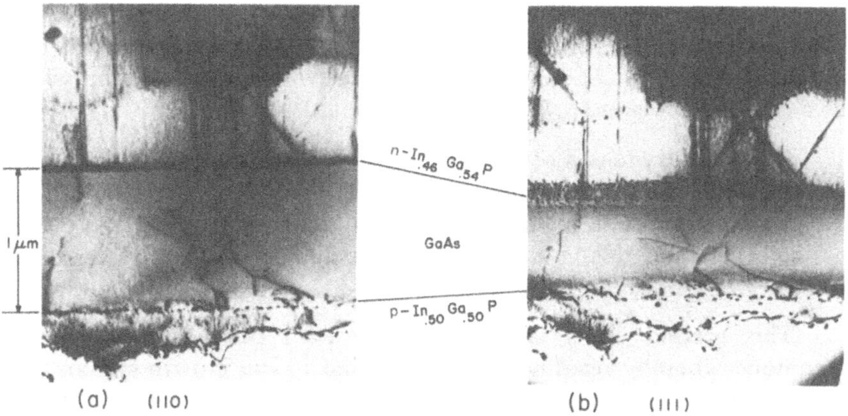

FIGURE 2. Bright-field cross-sectional transmission-electron microscope (XTEM) micrograph of the active GaAs region of a VPE double-heterojunction laser structure in (a) (001) and (b) (111) foil orientation. The extensive decomposition at the top interface (but not the bottom) is to be noted. Note also the large number of vertically inclined dislocations in the top (In,Ga)P layer (but not in the bottom) due to crack formation and subsequent propagation of dislocations down into the active GaAs region. (Courtesy of M. S. Abrahams and C. J. Buiocchi.)

before growth of the first p–$In_xGa_{1-x}P$ layer. After growth of the p–$In_xGa_{1-x}P$, the sample was withdrawn to the preheat (which then contained a PH_3 flow in addition to H_2) prior to deposition of the 1-μm GaAs active layer. After deposition of the 1-μm GaAs layer, the sample was again withdrawn to the preheat zone which contained only pure H_2. The n–$In_xGa_{1-x}P$ layer was then grown. It is evident that the top of the GaAs layer (for which there was no group V flow during the preheat) incurred substantial decomposition over a region ~500 Å thick, whereas the top of the p–$In_xGa_{1-x}P$ layer (which was preheated in PH_3 before growth of the GaAs) had little or no decomposition. Decomposition was also observed at the original GaAs/p–$In_xGa_{1-x}P$ interface. Thus, whenever a III–V compound is held at elevated (>600°C) temperatures before or between growth steps, it should be held in a proper atmosphere to prevent decomposition while the reactant flows in the growth zone are readjusted. To terminate growth, the sample is removed to the forechamber where it quickly cools to room temperature.

2.2.2. Composition Profiles at VPE Heteroepitaxial Interfaces

Since both the lattice parameter and bandgap vary directly with composition in III–V alloys, it is important to know how composition varies across a heteroepitaxial (e.g., GaAs/$In_xGa_{1-x}P$) interface. Figure 3 shows a secondary-ion mass spectrometer (SIMS) compositional profile obtained from a GaAs/$In_{0.5}Ga_{0.5}P$ interface, which was prepared via VPE as described previously. Two features should be noted:

1. A quaternary region of $In_xGa_{1-x}As_yP_{1-y}$ exists over an ~1000-Å-thick region at the interface.
2. A finite amount of indium (~3%) exists in the "GaAs" at the interface and this value tapers off much more slowly as compared to the phosphorus.

This second effect was due partly to a peculiarity of the VPE technique whereby residual InCl was found in the growth system during the growth of GaAs after $In_xGa_{1-x}P$, and probably partly to diffusion effects since In diffuses more rapidly in GaAs than either Ga, As, or P.[15] This affects the interfacial strain since the interface is no longer GaAs/$In_{0.5}Ga_{0.5}P$ but $In_xGa_{1-x}As/In_xGa_{1-x}P$. Figure 4 shows an XTEM of a supposedly lattice-matched GaAs/$In_{0.5}Ga_{0.5}P$ photocathode structure from which the data of Fig. 3 were taken. It is clear that the interface is not in fact lattice-matched and misfit dislocations are in evidence. The residual InCl can be reduced by appropriate flushing of the growth system

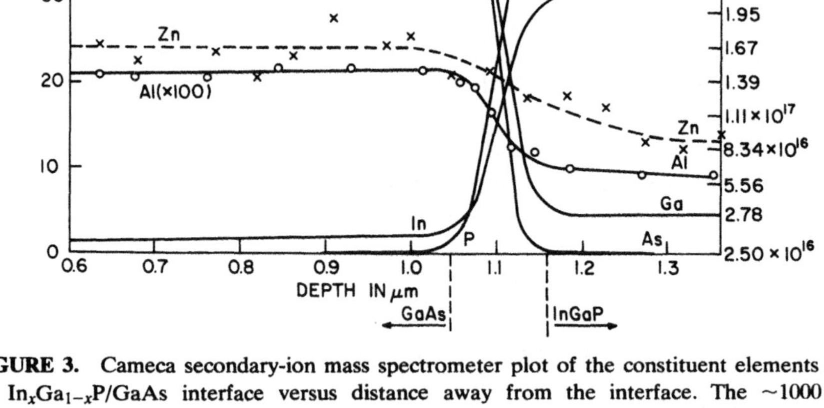

FIGURE 3. Cameca secondary-ion mass spectrometer plot of the constituent elements in an $In_xGa_{1-x}P$/GaAs interface versus distance away from the interface. The ~1000 Å quaternary region and the persistence of indium in the GaAs are to be noted. (Courtesy of W. L. Harrington.)

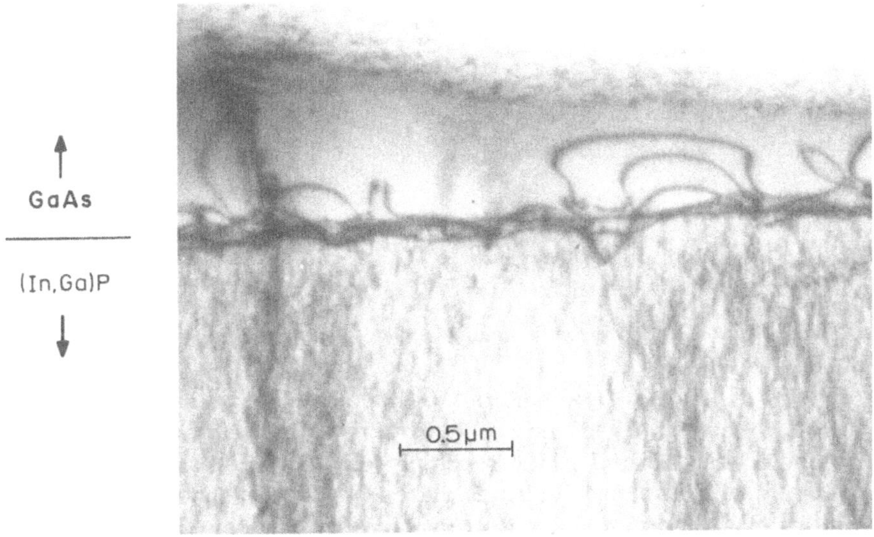

FIGURE 4. Bright-field {011} XTEM micrograph from a lattice-mismatched $GaAs/In_xGa_{1-x}P$ interface of a VPE transmission photocathode. The [100] growth axis is vertical. (Courtesy of the *Journal of Applied Physics.*)

between deposition of the two materials. However, the existence of a quaternary layer at the interface should be viewed as a fundamental aspect of heterojunctions, since significant diffusion of the III–V elements can occur at the elevated temperatures of growth.

2.3. Liquid-Phase Epitaxy (LPE) Growth Technique

Liquid-phase epitaxy (LPE) is the precipitation of material onto a substrate from solution. For the growth of III–V compounds, the solvent is the group III component, Ga and/or In with possible additions of Al, saturated with the group V component P, As, or Sb, or combinations of these. These solutions are used to form binary, ternary, and quaternary III–V semiconductors.

There are three basic LPE techniques: tipping,(16) dipping,(17) and slider boat.(18) For heteroepitaxy, where multiple layers are desired, the slider boat is most often used. Figure 5 is a photograph of one of the more sophisticated slider boats(19) with one side wall removed. This graphite boat will serve to illustrate the general growth technique, and is the boat used for much of the LPE growth described in this chapter. By using the growth of $Al_xGa_{1-x}As$ multilayers as an example, we describe the general technique.

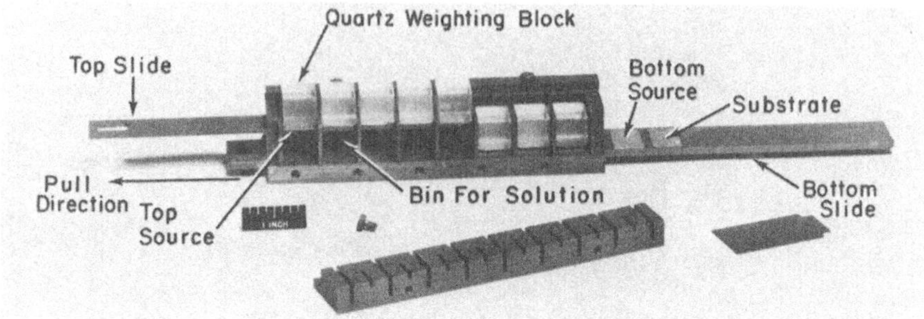

FIGURE 5. LPE thin solution multibin slider boat with one side wall removed.

For $Al_xGa_{1-x}As$ growth, the Ga solvent plus Al and dopants are added to each bin in amounts according to the Al concentration and doping density desired in the resulting solid. A GaAs source wafer used to saturate each melt with As is placed on the top slider in each bin as well as in the bottom slider in front of the substrate.

The boat is placed in a quartz tube which has a continuously flushed pure hydrogen atmosphere and the quartz tube is put into a furnace. After thermal equilibrium is attained at the desired saturation temperature, the top slide is removed and the source wafers and quartz weighting blocks are dropped in order to distribute the solution in each bin uniformly. After allowing sufficient time for complete saturation (~1 hr), the bottom source wafer is brought into the first bin, again to assure saturation, followed in a few minutes by the substrate. The furnace is then usually heated a few degrees to dissolve some of the substrate, and then cooled at about 0.5°C/min. After a temperature drop sufficient to grow the desired thickness in the first bin, the substrate is moved to the second bin to grow the second layer. The cooling proceeds in the second bin until the second layer is grown. The cooling is continuous and the substrate with its preceding source wafer proceeds from bin to bin until all the desired layers are grown. The substrate with epitaxial layers deposited is then wiped clean of solution in the last empty bin, and the tube is removed from the furnace.

The extra care taken to ensure that each of the solutions is saturated is derived from the desire to make LPE an equilibrium growth technique. In equilibrium, the thickness and composition of the grown layers are very often calculable from the appropriate phase diagram.[20] Also, if saturation were not carefully maintained, thin layers ($\leq 0.5\ \mu m$) grown in one bin could be dissolved in the next.

For graded structures, it should be clear from the preceding discussion that the bins of a slider boat are independent and abrupt changes in epitaxial layer composition are easily attained—for instance, uniform GaAs layers as thin as 400 to 500 Å have been bounded on both sides by $Al_xGa_{1-x}As$ using LPE.[21] However, continuous grading is very difficult and only really possible in systems in which the distribution coefficients from solution are favorable.[22]

3. ELASTIC STRAIN EFFECTS

3.1. Stresses in Heteroepitaxial Layers

In general, the presence of elastic strain between heteroepitaxial multilayers will have two effects:

1. A stress (σ_i) will exist in each layer.
2. The composite will bend (see Fig. 6) in response to these stresses.

The origin of these stresses might be differences in lattice parameter, thermal expansion coefficients, or other material properties (e.g., magnetostriction). It is often of interest to know the stress distribution and bending radius in such materials. The effect of stress on device performance in heterojunction lasers would be one example. It is also of great importance to know whether or not the critical yield stress of a material has been exceeded.

The problem is a difficult one to solve in general, and Hoffman[23] has given an excellent review of the various treatments that are available for stress calculations in crystals. Most of these calculations deal only with

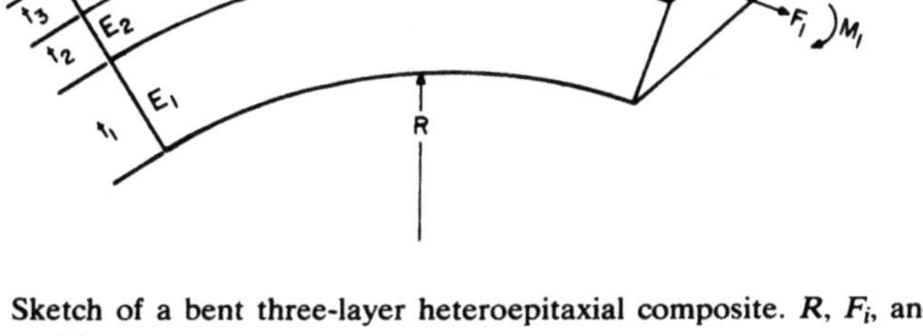

FIGURE 6. Sketch of a bent three-layer heteroepitaxial composite. R, F_i, and M_i are all drawn in the positive sense.

two-layer structures (a substrate and an overgrowth). Saul[24] has calculated the stress at the uppermost interface of an N-layer composite (e.g., Fig. 6), which is strained by differential thermal expansion.[90] The present authors[26] have extended these calculations so that the stress at any point in any layer of such a composite in which strains ϵ_{ij} exist between layers may be computed.

The solution consists of solving for the radius of curvature R and the N forces F_i. The stresses σ_i in each layer are then calculated in terms of these variables. The $N+1$ equations needed to solve for the forces and radius come from equilibrium of forces and moments (two equations) and continuity of length at each of the $N-1$ interfaces ($N-1$ equations).

It is shown in Appendix A that for structures such as that shown in Fig. 6 with $L \gg w$ and equal elastic constants, the radius of curvature R and stress σ_j in an N-layer heterostructure may be expressed in closed form as

$$R = t^3 \left[6 \sum_{i=1}^{N} \sum_{j>i}^{N} t_i t_j \epsilon_{ij} \right]^{-1} \tag{3}$$

$$\sigma_j(y) = E \left[\sum_{i=1}^{N} \left(\frac{t_i \epsilon_{ij}}{t} + \frac{t_i \kappa_i}{2R} \right) + \frac{1}{2R} \left(y - \frac{t_j}{2} \right) \right] \tag{4}$$

where t_i is the thickness of the ith layer;

$$t = \sum_{i=1}^{N} t_i$$

y is the distance from the *bottom* of the jth layer; and

$$\kappa = \begin{cases} +1, & i<j \\ 0, & i=j \\ -1, & i>j \end{cases}$$

The expression for R is well known and was first derived by Timoshenko.[25] The assumption of $L \gg w$ means that the solution is one-dimensional and neglects any effects of the transverse dimension. However, many device structures have this geometry and, furthermore, reasonable agreement (within 20%) has often been observed experimentally[26] between radii of curvature measured on long strips and square pieces of the same III–V heterostructure. The assumption of equal elastic constants is often a reasonable one for III–V compounds, although variations of 20% or more do occur.[27] In the examples considered here (e.g., Fig. 11), the error incurred by assuming equal elastic constants was roughly equal in magnitude to the difference in elastic constants. Further details may be found in a forthcoming paper.[26] In any case, the computer solution (see Appendix A) may be used when equal elastic constants are not assumed. Thus, Eqs. (3) and (4) are thought to be useful expressions for calculating stress in many multilayer III–V heterostructures. However, the expressions are also of use for any heterostructure in which the strains ϵ_{ij} may be defined. Even if interfacial misfit dislocations are present, the calculation is still valid as long as the residual elastic strain (if any) in the layers can be specified. Conversely, if the thicknesses and radius of curvature can be measured, the amount of residual elastic strain present can be calculated. The presence of inclined dislocations in the layer, on the other hand, has a complicating effect on the bending which is discussed elsewhere.[28] Several examples are now presented to demonstrate the usefulness of Eqs. (3) and (4).

One of the more interesting systems for stress calculations can be found in heteroepitaxial composites of LPE $Al_xGa_{1-x}As$/GaAs. This system is of interest because (a) it has been employed to produce a wide variety of optoelectronic devices[29] whose properties are unmatched by other material systems; and (b) no misfit dislocations are observed in this system. The lack of misfit dislocations might at first seem surprising, since the room temperature lattice mismatch between AlAs and GaAs is 1.26×10^{-3}.[30] However, due to differences in thermal expansion coefficient, the lattice parameters of the two materials (see Fig. 7) are virtually identical[30] at the growth temperature (800–900°C). Thus, during the

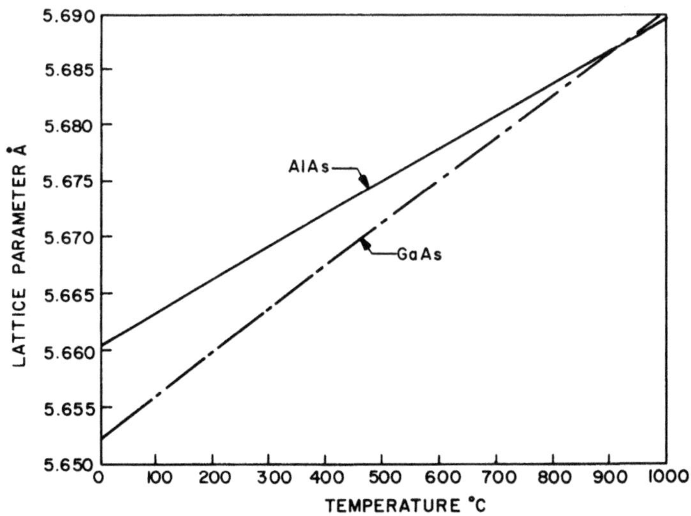

FIGURE 7. Lattice parameters of GaAs and AlAs as a function of temperature. (Courtesy of the *Journal of Applied Physics.*)

growth process, no lattice mismatch exists, and, therefore, no misfit dislocations are generated. Only when growth is terminated and the wafer cooled is strain introduced. However, since plastic deformation (slip) is unlikely[31] below 500°C, and the strain at this temperature is only $\sim 5 \times 10^{-4}$, the critical resolved shear stress for this material is at most only marginally exceeded for a few seconds during the cooling process after growth. Furthermore, climb processes,[32] which might also introduce dislocations, do not have sufficient time to occur.

Figure 8 is a reflection X-ray topograph of an LPE-grown 2-μm-thick layer of AlAs (Ga < 2%) on a 250-μm-thick low dislocation density ($<1 \times 10^3\ \text{cm}^{-2}$) GaAs substrate. Except for a few scratches which appear to propagate from the substrate, the AlAs layer is free of dislocations. Calculations show that the room temperature stress in the AlAs is $1.25 \times 10^9\ \text{dyn/cm}^2$ and that the radius of curvature is 4.3 m, which is in good agreement with the measured radius of 4.0 m. Because of the absence of strain-relieving dislocations, the $Al_xGa_{1-x}As$ system is amenable to the analysis just derived for conditions of simple strain.

One of the more interesting structures comprising heteroepitaxial layers of $Al_xGa_{1-x}As$ is the double-heterojunction cw laser. The typical cw laser is a five-layer structure[33] comprising a substrate and four epitaxial layers of varying thickness and Al concentration. A typical cw structure is illustrated in Fig. 9. Using the equations developed in the preceding

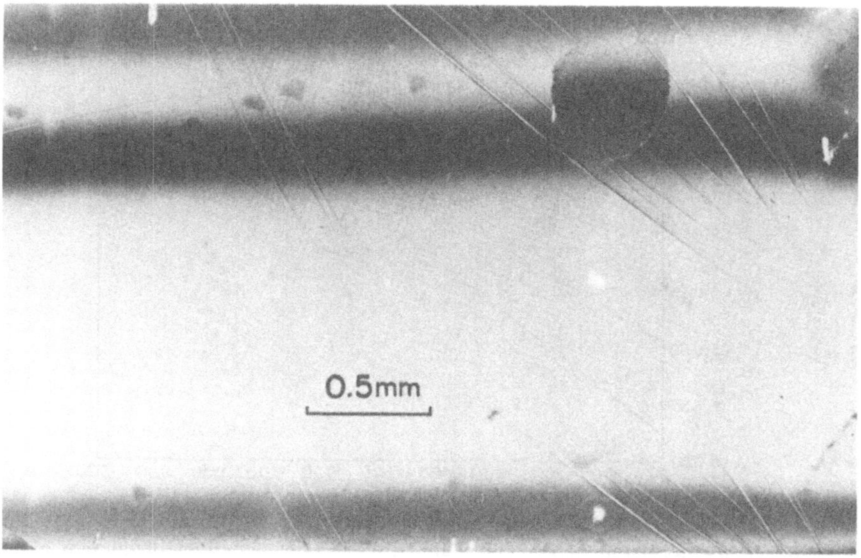

FIGURE 8. Reflection X-ray topograph of a (111) LPE AlAs/GaAs showing the absence of interfacial misfit dislocations (compare with Fig. 13). (Courtesy of S. H. McFarlane.)

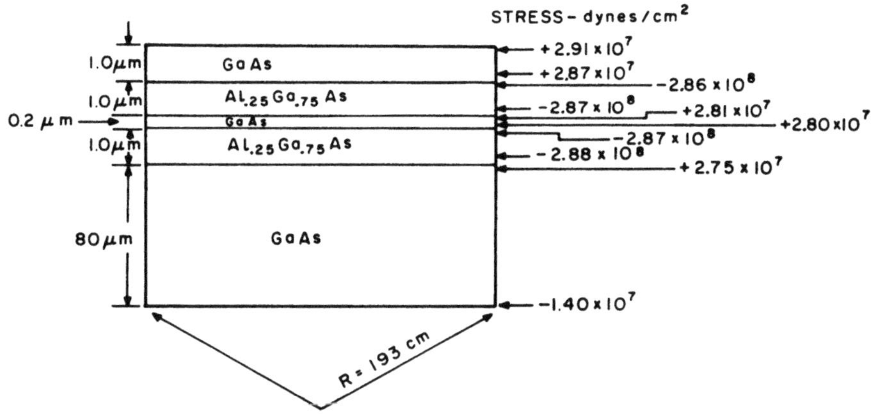

FIGURE 9. Schematic cross section of a typical LPE GaAs cw laser with the calculated radius of curvature and stress in each layer.

section, the stress levels at each interface and the radius of curvature have been calculated. The conventions chosen are illustrated in Fig. 6 for an elastically bent homogeneous solid. Positive (+) is tension, and negative (−) is a state of compression. The radius of curvature is positive (+) if measured from the bottom, which is always substrate (concave down). In these calculations, we have assumed that Young's modulus is 1×10^{12} dyn/cm^2 for the entire $Al_xGa_{1-x}As$ alloy system,[34] and that the room temperature lattice mismatch is 1.26×10^{-3} between AlAs and GaAs and linearly related to the AlAs concentration.[30]

The qualitative features of the stress levels in the $Al_xGa_{1-x}As$ heterojunction structure can be easily understood once one realizes that $Al_xGa_{1-x}As$ has an unconstrained room temperature lattice parameter larger than that of GaAs and that the GaAs substrate, because it is much thicker than all the epitaxial layers combined, dominates the structure. Thus, the $Al_xGa_{1-x}As$ epitaxial layers are in a state of compression. To relieve the overall stress, the composite bows so that the $Al_xGa_{1-x}As$ layers on the top can expand. This puts the bottom of the substrate in compression and the epitaxial GaAs near the top, including the GaAs active region between the $Al_xGa_{1-x}As$ layers, in a state of tension.

One might investigate the effect of various parameter changes, such as Al concentration and layer thickness, on the stress levels in the various layers. Since in most useful device structures the epitaxial layers are relatively thin compared to the substrate, it is the substrate which dominates; reasonable changes in epitaxial layer thickness change the qualitative picture not at all, and the quantitative aspects in a relatively linear fashion.

Table 2. Stress Levels in Double-Heterojunction Laser Structure

		GaAs substrate		$Al_{0.25}Ga_{0.75}As$		GaAs active layer		$Al_{0.25}Ga_{0.75}As$		GaAs "cap"	
Structure	R (m)	Thickness (μm)	Stress ($\times 10^{-8}$ dyn/cm^2)	Thickness (μm)	Stress ($\times 10^{-5}$ dyn/cm^2)	Thickness (μm)	Stress ($\times 10^{-8}$ dyn/cm^2)	Thickness (μm)	Stress ($\times 10^{-8}$ dyn/cm^2)	Thickness (μm)	Stress ($\times 10^{-8}$ dyn/cm^2)
1	1.9	80	0.28	1.0	−2.9	0.2	0.23	1.0	−2.9	1.0	0.29
2	0.55	40	0.48	1.0	−2.7	0.2	0.50	1.0	−2.6	1.0	0.52
3	7.2	160	0.15	1.0	−3.0	0.2	0.15	1.0	−3.0	1.0	0.15
4	1.0	80	0.51	2.0	−2.6	0.2	0.53	1.0	−2.6	1.0	0.56
5	2.0	80	0.26	1.0	−2.8	0.2	0.27	1.0	−2.8	2.0	0.27
6	1.9	80	0.28	1.0	−2.9	0.1	0.28	1.0	−2.9	1.0	0.29
7	1.8	80	0.29	1.0	−2.9	0.2	0.29	1.0	−2.9	0.1	—

Table 2 lists calculated values of stress level in each epitaxial layer and at the top of the substrate for a variety of useful layer thicknesses in the general cw laser structure illustrated in Fig. 9. For the relatively thin epitaxial layers, there is little variation in the stress level from top to bottom of each layer (as indicated in Fig. 10); the stress at the top of the layer is reported in this table.

The first aspect of these results to note is that the stress is compressive in the $(Al_{0.25}Ga_{0.75})As$ confining layers and is relatively independent of layer thickness in the limited range studied. In a laser structure, the thin GaAs layer confined between the two $Al_xGa_{1-x}As$ heterojunctions is called the active region and is probably the most important region, since the electron–hole recombination and the lasing emission are confined to this layer. If we study this region, we see that the thicker the substrate, the lower is the stress in the active region. One might expect this result, since a thicker substrate constrains the $Al_{0.25}Ga_{0.75}As$ layers closer to the GaAs lattice parameter, and thus the active GaAs layer will be closer to its normal lattice parameter. This effect is observed in structures 1, 2, and 3. Thickening the $Al_{0.25}Ga_{0.75}As$ layers, as in structure 4, increases the stress in the active layers and thus

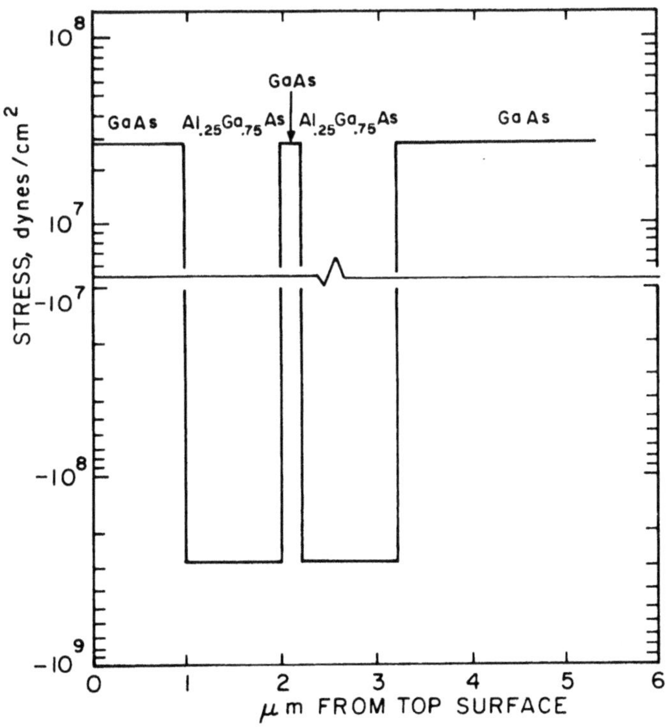

FIGURE 10. Stress as a function of thickness in the structure of Fig. 9.

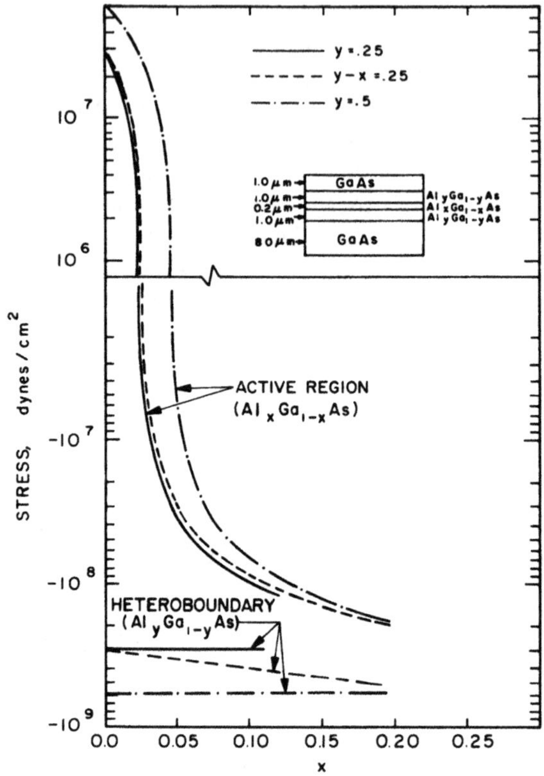

FIGURE 11. Stress in the active region and in the heteroboundary layers as a function of Al concentration for various useful differences in Al concentration at the heterojunctions. The structure considered is that illustrated in Fig. 9.

has a similar effect to thinning the substrate. Again, because the GaAs substrate is so thick, thinning the GaAs active layer, as in structure 6, or completely removing the GaAs "cap" layer, as in structure 7, has almost no effect on the stress level in any of the layers.

The most dramatic effect, especially on the stress in the active region, is due to changes in the active region Al concentration. Plotted in Fig. 11 is the active region stress as a function of active region Al concentration x for the thicknesses illustrated in Fig. 9. Three different regimes are studied: where the Al concentration in the confining layers (y) is constant at 25% ($y = 0.25$), where they are constant at 50% ($y = 0.5$), and where the difference between the active region and confining layer Al concentration is constant at 25% ($y - x = 0.25$). In each of these regimes, the active region stress is reversed from a state of tension to one of compression with small additions of Al. The crossover from tension to compression occurs at about 5% Al ($x \simeq 0.05$), independent of the Al concentration at the heteroboundaries. As the Al concentration of the active region is increased, the active region stress approaches that of the confining $Al_yGa_{1-y}As$ layers. The stress in the confining layers for the three regimes is also plotted in Fig. 11.

The variations of stress with layer thickness calculated for $Al_xGa_{1-x}As/GaAs$ lasers would be the same for VPE $In_{0.5}Ga_{0.5}P/GaAs$ lasers. However, the composition (x) of the confining $In_xGa_{1-x}P$ layer must be fixed rigidly at $x = 0.504 \pm 0.002$ in order to avoid misfit dislocation introduction. Thus, one obvious disadvantage with the use of (In,Ga)P confining layers is the need for very precise composition control during fabrication. However, it does have the advantage that its composition can be adjusted to match the lattice parameter of $In_yGa_{1-y}As$ active regions for $0 < y < 30.53$ (i.e., emission wavelengths as high as 1.6 μm[35]).

The effect of stress on the electronic properties of III–V compounds has not been clearly defined. It has been shown to alter the bandgap slightly[36] and to affect the degradation rate in LEDs[37] and cw lasers.[38] It may also affect the interfacial recombination velocity at heterojunctions.[39] A bandgap shift of 15 meV has been measured via photoluminescence in a lattice-mismatched VPE layer of $In_{0.52}Ga_{0.48}P$/GaAs both before and after removal of the GaAs substrate.[40] This shift is reasonably consistent with the reported[41] increase of bandgap with hydrostatic pressure (11 meV/kbar) in GaAs. With the ability to calculate the stress level in various III–V heterostructures, it may be possible to quantify the effect of stress and strain on the electronic properties of heteroepitaxial devices. For instance, the stress gradient ($\sim 1 \times 10^5$ dyn/cm^2) across the active region of the cw laser structure shown in Fig. 9 corresponds to a strain gradient of $\sim 5 \times 10^{-7}\ \mu m^{-1}$. This value is almost two orders of magnitude lower than the value found to accelerate degradation in lasers by Johnston and Logan.[42]

3.2. Tetragonal Distortion

X-ray analysis has revealed a tetragonal distortion of the crystal structure in lattice-mismatched VPE and LPE materials.[43,44] The distortion comes about from the strain being applied in the interfacial plane while the direction normal to it (c) is free to expand or contract due to the Poisson effect, as illustrated in Fig. 12. An example of the effect in VPE materials has been demonstrated[44] in (001) $In_xGa_{1-x}As/GaAs$. Lattice parameters were measured in a series of four step-graded $In_xGa_{1-x}As$/GaAs crystals, both with the epitaxial layer on the substrate (single-crystal diffractometer) and with the epitaxial layer scraped off the substrate (Debye–Scherrer powder technique). The crystals were graded in compositional steps of 2, 4.8, 10.4, and 20.5% In, with each one of

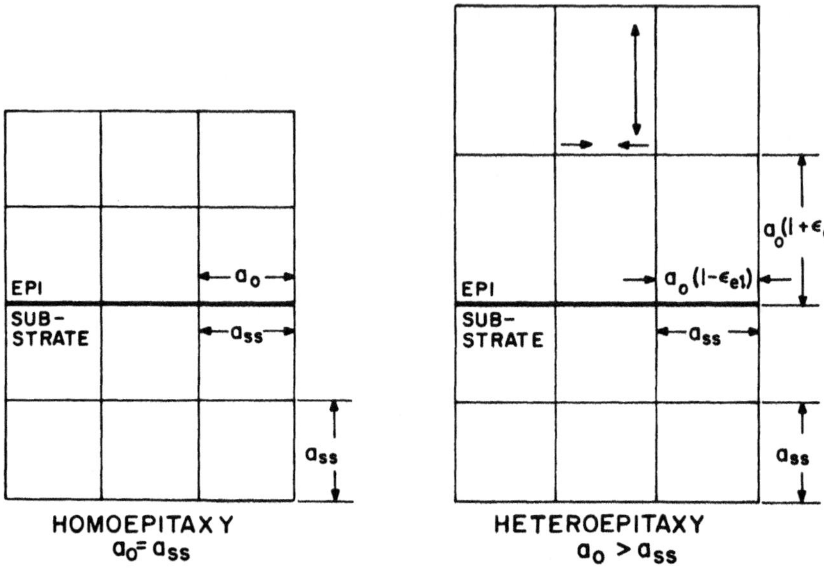

FIGURE 12. Tetragonal distortion for a coherent epitaxial layer having a larger lattice parameter than its substrate.

these compositions being that of the final layer, as shown in Table 3. All symbols are as defined in the table. The value c is the diffractometer lattice parameter of (001) planes along the substrate axis. The value a_{pow} is the powder pattern lattice parameter and is assumed to represent the actual, unstrained lattice parameter of the epitaxial layer from which the chemical composition may be derived via Vegard's law. The value a is the strained lattice parameter along the two <100> directions that lie in the substrate. It is shown in Appendix B that these values are related by

$$a = a_{pow}\left(2 - \frac{c}{a_{pow}}\right) \tag{5}$$

The values listed in Table 3 were then calculated. The c/a ratio is seen to increase with the lattice mismatch in the final layer until a value of at least 10% In is reached. The ratio then appears to decrease for larger values of misfit. The ratio of elastic strain to total misfit strain in the final layer, on the other hand, remains constant until the misfit exceeds 4×10^{-3}, when it too drops off. These results indicate that until the lattice misfit exceeds 4×10^{-3}, only about 60% of this misfit is accommodated by misfit relief mechanisms (e.g., dislocations). The rest of this misfit gives rise to an elastic strain, which, in turn, accounts for the increasing Poisson extension normal to the substrate. Once the misfit reaches 7.2×10^{-3}, however, misfit dislocations have been introduced that relieve over 90% of the misfit strain and thus lower the c/a ratio.

The tetragonal distortion was also measured directly via single-crystal diffractometry in two $In_xGa_{1-x}As/GaAs$ crystals. The distortion was established by taking the ratio of lattice parameter measured along the [001] substrate axis (c) to that along the ⟨100⟩ direction (a) in the substrate as measured on adjacent ⟨110⟩ cleavage edges. A c/a ratio of 1.0016 was measured in an abrupt (ungraded) layer of $In_{0.15}Ga_{0.85}As/GaAs$, which is consistent with Nagai's results[43] (which presumably were also obtained with ungraded samples). In a step-graded layer of $In_{0.2}Ga_{0.8}As/GaAs$, a c/a ratio of only 1.0007 was observed in the uppermost layer. However, the compositional step between the last and preceding layers was only 5%, so the result is actually consistent with Nagai's data and shows that the tetragonal distortion also varies with grading technique.

Nagai has measured tetragonal distortions in VPE layers of $In_xGa_{1-x}As/GaAs$ with c/a ratios as high as 1.0026, although the grading technique (if any) was not specified. We have measured similar values in step-graded VPE layers of $In_{0.2}Ga_{0.8}As/GaAs$ whereby the composition was changed in abrupt steps from $x = 0$ to $x = 0.2$. Nagai proposes that the tetragonal distortion is due to incomplete misfit relief by misfit dislocations which would leave an elastic strain (e.g., "bowing of the wafer") present in the crystal. The evidence presented here agrees with this conclusion and demonstrates that about 40% of the total lattice misfit is accommodated by elastic strain until a misfit $>4 \times 10^{-3}$ is achieved. This suggests that (a) an insufficient number of dislocation sources is available, or (b) there is insufficient stress to move them. The second condition could also occur if the elastic strain dropped below a critical value once a certain number of misfit dislocations had been introduced. The present

Table 3. Tetragonal Distortion in $In_xGa_{1-x}As$ *Solid Solutions*

Parameter[a]	In content (atm %)			
	2	4.8	10.4	20.5
c	5.6640	5.6775	5.7040	5.7390
a_{pow}	5.6614	5.6728	5.6957	5.7362
a	5.6587	5.6683	5.6872	5.7333
c/a	1.0009	1.0016	1.0030	1.0010
ϵ_{el}	0.0005	0.0008	0.0015	0.0005
ϵ_{fl}	0.0014	0.0020	0.0040	0.0072
$\epsilon_{el}/\epsilon_{fl}$	0.36	0.40	0.38	0.07
$\epsilon_{mf(total)}$	0.0014	0.0034	0.0075	0.015

[a]Here c is the lattice constant along the c axis [100] (normal to wafer) measured by single-crystal diffractometry; a_{pow} is the lattice constant measured by the Debye–Scherrer powder technique; a is the lattice constant along the [010] and [001] directions (lying in the wafer) calculated from $a = a_{pow}(2 - c/a_{pow})$; ϵ_{el} is the unrelieved elastic strain $\simeq c/a_{pow} - 1$; ϵ_{fl} is the total mismatch between the final layer and the next to final layer; and $\epsilon_{mf(total)}$ is the total mismatch between the final layer and the GaAs substrate.

results indicate that this critical strain is $\approx 4 \times 10^{-3}$. Interestingly, the critical value of misfit strain necessary for the onset of cracking (which also involves dislocation motion) in deposits of $In_xGa_{1-x}P/GaAs$ (in tension) was found to be $\approx 3 \times 10^{-3}$.[10]

Tetragonal distortions of the same order were also observed in deposits of $In_xGa_{1-x}P$ and $GaAs_{1-x}P_x$ on GaAs and GaP substrates, and for both step-graded and continuously graded layers. [Layers in tension, e.g., Ga(As,P)/GaAs, exhibit $c/a < 1$.] It should also be added that all of these crystals exhibited both misorientation and tetragonal distortion.

Tetragonal distortion has also been observed in $Al_xGa_{1-x}As$ grown on GaAs by LPE.[70] Our results for the LPE deposit of AlAs on GaAs in Fig. 8 show a lattice constant of 5.667 Å for planes parallel to the substrate plane. Assuming the room temperature mismatch to be completely accommodated by elastic strain, Eq. (5) predicts a value of 5.668 Å compared to 5.661 Å for the undistorted AlAs lattice. This agreement again demonstrates the Poisson-induced tetragonal distortion and serves as further evidence that the misfit between AlAs and GaAs grown via LPE is totally accommodated by elastic strain rather than misfit dislocations.

3.3. Coherency Limits in III–V Compounds

In principle, coherent growth of any material can be achieved for thicknesses less than that predicted by Eq. (2). Coherency here is taken to mean that the overgrowth lattice parameter is constrained to match that of the substrate. The extent to which experimental and theoretical values of h_c agree depends strongly on growth conditions. The following examples will illustrate this point.

Figure 13 depicts an X-ray topograph of a 2-μm layer of VPE $GaAs_{0.99}P_{0.01}$ on a GaAs substrate. The misfit strain (ϵ) for this composite is $\sim 4 \times 10^{-4}$ and Eq. (2) predicts $h_c \sim 2500$ Å. Misfit dislocations are thus expected to be present and Fig. 13 indeed shows this to be true. Since no curvature could be detected via X-ray topography, all of the misfit strain should be accommodated by misfit dislocations. (The actual dislocation density could not be ascertained from Fig. 13. However, other workers[28] have shown this system to contain near-equilibrium misfit dislocation morphologies.) Experimental and theoretical observations thus seem consistent for this system.

Figure 8 depicts an X-ray topograph of a 2-μm layer of LPE AlAs on a GaAs substrate. No misfit dislocations are evident even though the room temperature misfit is 1.26×10^{-3} and $h_c \sim 1600$ Å. A radius of curvature of 4.0 m was measured, which agrees well with a predicted

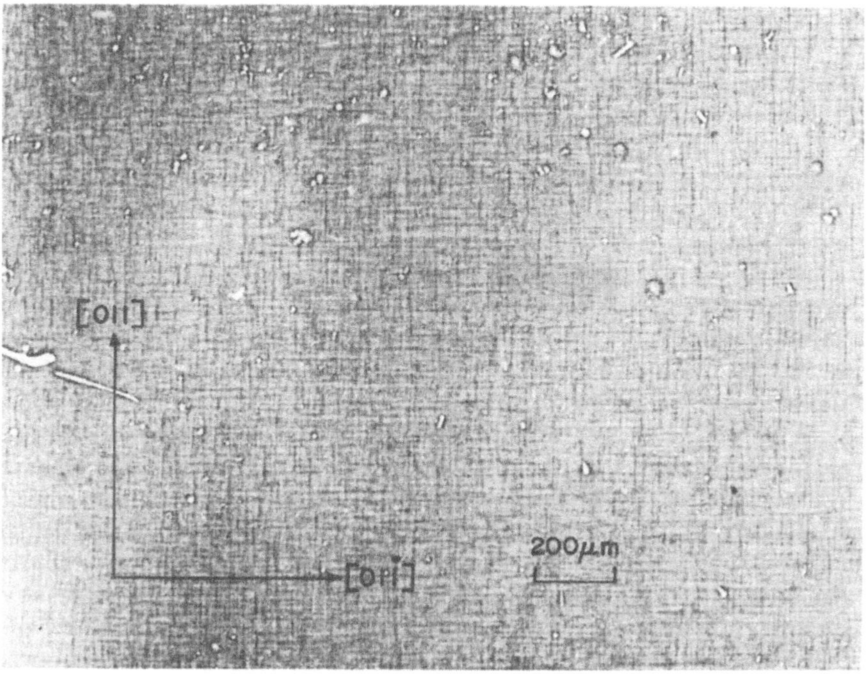

FIGURE 13. Reflection X-ray topograph of a VPE $GaAs_{0.99}P_{0.01}$ layer on a (100) GaAs substrate showing misfit dislocations. (Courtesy of S. H. McFarlane.)

value (see Section 3.1) of 4.3 m. All of the misfit strain is therefore accommodated elastically even though the room temperature strain is three times higher than that of the $GaAs_{0.99}P_{0.01}$/GaAs sample which had plastic deformation. However, this difference has already been explained as a thermal expansion effect ($\epsilon \sim 0$ at the growth temperature of LPE AlAs). Rozgonyi *et al.*[45] have taken advantage of this behavior and have eliminated both elastic strain *and* misfit dislocations in laser heterostructures by introducing small amounts of phosphorus into $Al_xGa_{1-x}As$. In the LPE growth of $Al_xGa_{1-x}P_yAs_{1-y}$ on GaAs, the value of y can be controlled such that a misfit strain exists at the growth temperature but disappears at room temperature. If the thickness of the layer is kept below the critical thickness for the misfit strain at the growth temperature (which in this case can be on the order of micrometers), the final layer is free of both elastic strain and dislocations at room temperature. Rozgonyi *et al.*[45] and Petroff[46] have made an interesting study of the variation of h_c with phosphorus content in $Al_xGa_{1-x}P_yAs_{1-y}$/GaAs. The value of h_c was determined by X-ray topography and transmission electron microscopy by determining which alloys contained misfit dislocations. The value of misfit strain was then deduced from Matthews'[5] expression for the variation of h_c with misfit. The values obtained seem reasonable and show that coherency is observed in 1-μm films for $\epsilon_{mf} \leqslant 1 \times 10^{-4}$. Asymmetric dislocation introduction was also observed in this study.

Figure 14 shows X-ray topographs and SEM cathodoluminescence photographs from different regions of a 2-μm VPE layer of $In_xGa_{1-x}P$/GaAs in which x varied across the layer from 0.46 (in which the layer was under tension) to 0.53 (compression). It must be emphasized that the compositional variation was in the plane of the wafer—*not* in the growth direction. Several interesting features are evident in these figures. Normal misfit dislocation configurations are observed on both the tension ($\epsilon \sim 2 \times 10^{-3}$) and compression ($\epsilon \sim -3 \times 10^{-3}$) ends. [In SEM cathodoluminescence, dislocations near the surface (within $\sim 1\ \mu$m) appear black because they are centers for nonradiative recombination.[47]] Near the middle of the wafer ($x \sim 0.50$, $\epsilon \sim 0$) no misfit dislocations are observed over a narrow range. This region defines the crossover from tension to compression and also serves to define the limits of coherency for a 2-μm VPE $In_xGa_{1-x}P$/GaAs composite. Since these two materials also have different thermal expansion coefficients,[10] the "zero" of strain at the growth temperature (700°C) occurs at a different value of x ($\sim$0.495) than at room temperature ($x \simeq 0.483$). The misfit strain between

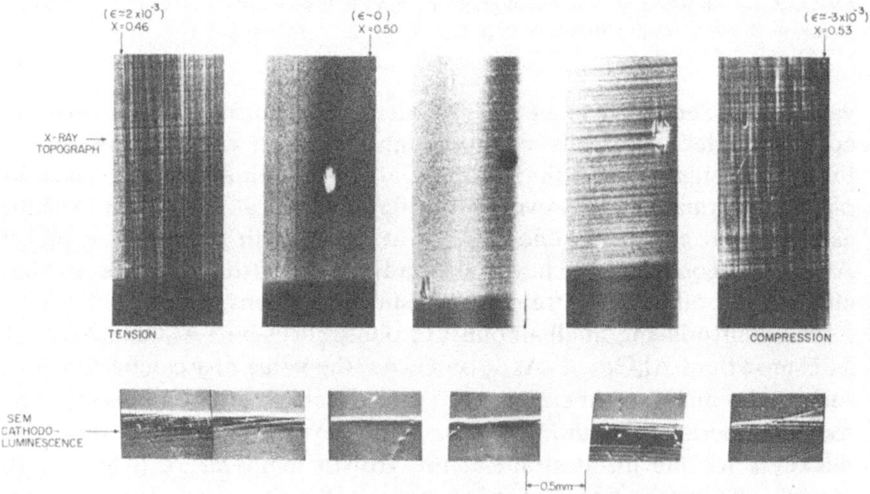

FIGURE 14. Reflection X-ray topographs and SEM cathodoluminescence photographs from various regions of a VPE $In_xGa_{1-x}P$ layer on a (100) GaAs substrate in which x varied as shown. Note the misfit dislocation distribution as a function of composition. The cathodoluminescence photographs were taken after the sample was angle-lapped in order to reveal the interface. (Courtesy of S. H. McFarlane and E. Levin.)

these two temperatures is $\sim 9 \times 10^{-4}$. Thus, it is difficult to define an "unstrained" region in terms of composition. All that can be said is that the misfit strain in the dislocation-free region was less than the coherency limit at the growth temperature. The compositional difference between the two ends of the dislocation-free region was measured via photoluminescence[40] to be $x = 0.502$ to 0.506, equivalent to a lattice parameter difference of $\sim 3 \times 10^{-4}$. Thus, for a 2-μm $In_xGa_{1-x}P$/GaAs, the coherency limit would correspond to a strain of $\sim 1.5 \times 10^{-4}$, which is consistent with the value observed by Rozgonyi *et al.*[45] for (Al,Ga)(As,P) alloys.

4. DISLOCATION EFFECTS

Whenever the conditions described in the preceding section for coherent growth of heteroepitaxial layers are exceeded, misfit dislocations will be generated to accommodate the lattice mismatch. Thus, for example, a 1-μm epitaxial layer having a misfit of 3% with its substrate at the growth temperature would certainly be expected to contain misfit dislocations. However, the distribution of these dislocations and the resultant perfection of the epitaxial layer are largely controlled by crystal growth techniques. These techniques are now described for both vapor-phase and liquid-phase epitaxy, and electron microscope and X-ray evidence from the resultant microstructures of such materials is presented.

4.1. VPE Materials

4.1.1. Misfit Dislocation Morphology Due to a Compositional Step

Cross-sectional transmission-electron microscopy (XTEM) provides a powerful technique for examining crystal defects and relating them to growth phenomena, since the growth axis is contained in the plane of observation and mechanical defects can be located precisely with respect to layer thickness.[48] Abrahams and Buiocchi[49,50] have pioneered in this field and have provided a great deal of direct experimental evidence on the introduction and morphology of misfit dislocations in III–V compounds. Figure 15 shows a series of XTEM micrographs of $In_xGa_{1-x}P$/GaAs interfaces having various amounts of abrupt (nongraded) lattice mismatch. These micrographs and many similar ones on other III–V systems[51,52] show that misfit dislocations in III–V compounds are predominantly of the 60° type lying along ⟨110⟩ directions with ⟨011⟩

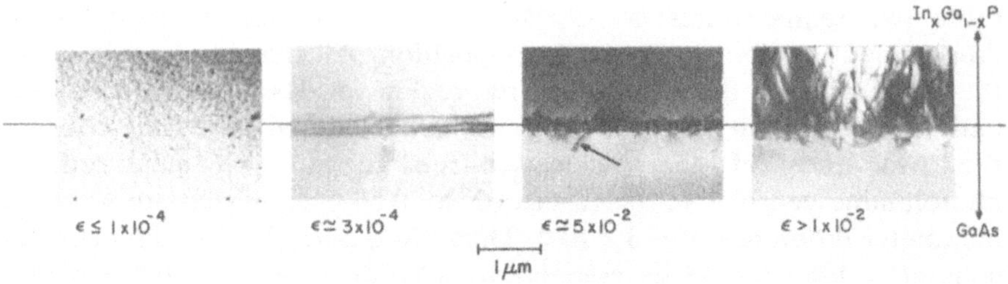

FIGURE 15. {011} XTEM micrographs from VPE $In_xGa_{1-x}P/GaAs$ interface having mismatch (ϵ) as shown. The [100] growth axis is vertical. The arrow indicates a substrate dislocation which has been bent over at the interface due to the lattice mismatch. (Courtesy of M. S. Abrahams and C. J. Buiocchi.)

Burger's vectors. From these photographs, and many similar ones, the following relationships between strain and misfit dislocations may be observed:

1. $\epsilon \leqslant 1\times10^{-4}$: No misfit dislocations are introduced.
2. $\epsilon \simeq 2\times10^{-4}$: Misfit dislocations are observed along only one ⟨011⟩ direction.
3. $3\times10^{-4} < \epsilon < 1\times10^{-2}$: Orthogonal arrays of misfit dislocations are observed at the interface. However, they are confined to within a few thousand angstroms of the interfacial region and do not propagate along the growth axis.
4. $\epsilon > 1\times10^{-2}$: Dense arrays ($>10^{10}\ cm^{-2}$) of dislocations which propagate away from the interface along the growth direction are observed.

The first observation is consistent with Eq. (2), which predicts that the film thickness should be $\geqslant 2\ \mu m$ for dislocation introduction. The second observation provides evidence for asymmetric dislocation introduction which was also seen in Fig. 14. The third observation shows that, within the specified range of strain, dislocations are either introduced along the interfacial plane or they get bent over into this plane as described by Matthews,(5) whereby they relieve the most strain. The final observation indicates that compositional steps which contain 1% or more misfit strain should be avoided in VPE materials. The large density of inclined dislocations probably results from misfit dislocations being introduced at the island stage of growth before a continuous film has formed. (Abrahams(53) has found that VPE layers coalesce into a continuous film at ~1000 Å, which is much greater than h_c for a 1% misfit.) If the misfit dislocations present in each island were to remain in the interfacial plane after coalescence, the overall dislocation density would increase manyfold unless an exact one-to-one matchup of dislocations occurred at each island boundary (which is highly unlikely).

Since the dislocation density needed to relieve the misfit would be exceeded in this case, a more plausible happening would be for the dislocations to incline away from the interface and propagate along the growth axis, as is seen in Fig. 15.

4.1.2. Sources of Misfit Dislocations

The fundamental creation and introduction of misfit dislocations has been a subject of perennial concern and much experimental evidence is available. Pashley[7,54] lists the following as possible sources of dislocations in epitaxial growth processes: (a) extension of substrate dislocations; (b) accommodation of misorientation of coalescing nuclei at early stages of growth; (c) dislocation half-loops formed by point defects; and (d) plastic deformation of the film. The bulk of the experimental evidence indicates that while the predominant method depends mostly on experimental conditions, method (c) has been observed the least. Figure 15 shows an example of the bending of an inclined substrate dislocation into the interfacial plane to accommodate misfit at a compositional step in VPE (In,Ga)P which had a misfit strain of $\sim 3 \times 10^{-3}$. Substrate dislocations will, in general, propagate into the epitaxial layer unless they are somehow bent over [e.g., by misfit strain(s)]. Dislocation motion (i.e., plastic deformation) has been observed[31] in III–V compounds at stresses (σ) as low as 1×10^8 dyn/cm^2 at temperatures $\geqslant 500°C$. This suggests that strains (ϵ) as low as 1×10^{-4} (assuming a Young's modulus[34] of 1×10^{12} dyn/cm^2 and $\epsilon = E/\sigma$) would, in principle, be able to cause dislocations to move into the interfacial plane during growth.

This lower limit of strain is also consistent with Rozgonyi's observations[45] for the bending in of substrate dislocations in LPE $Al_xGa_{1-x}As_yP_{1-y}$. An example of dislocation introduction via plastic deformation[50] may be seen in Fig. 16. Dislocations appear to have been introduced via slip at some distance above the interface and then to have glided down toward the compositional step. Dislocations that have a misfit-relieving Burger's vector are either stopped at the interface or move along it by cross-slipping on an adjacent {111} plane. Once dislocations have been introduced, multiplicative mechanisms can then operate to increase their density. Figure 4 is a dramatic illustration of a classic Frank–Read[55] source in action at a mismatched GaAs/$In_xGa_{1-x}P$ interface.[50] Dislocations at the interface bow out due to the misfit stress, but get pinned due to interaction with other dislocations. If the stress is sufficiently high, the dislocation will bow out between the pinning points until it pinches itself off and forms a closed loop (leaving the original dislocation behind). Because the loop has the same Burger's vector along

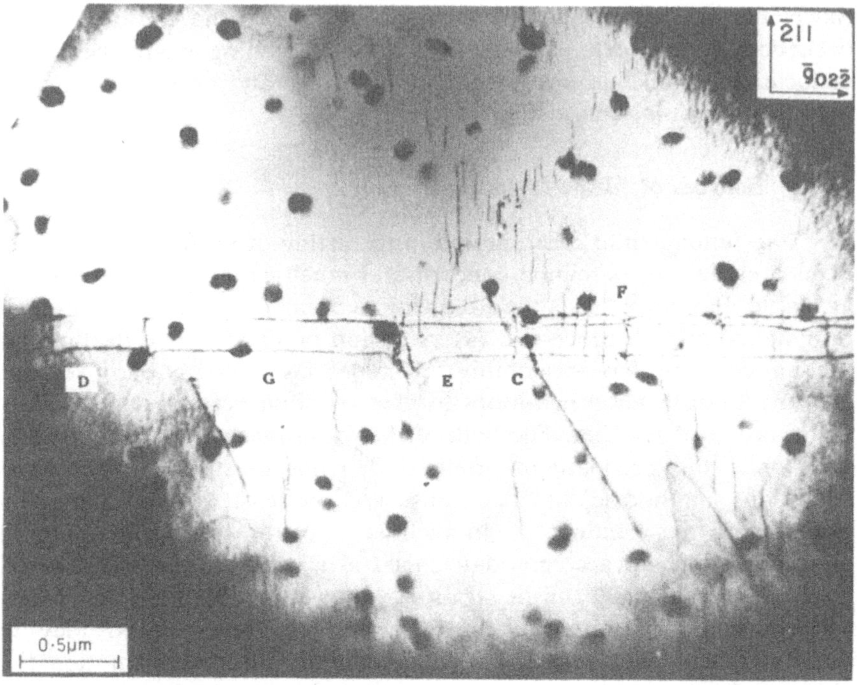

FIGURE 16. XTEM of a compositional step in a VPE $In_xGa_{1-x}P$ crystal in {111} foil orientation in which misfit dislocations at the step have originated from slip which took place further up along the growth axis. (Courtesy of the *Journal of Applied Physics.*)

its length, part of it would be able to relieve misfit and remain in the interface while the rest would incline away from the interface.

4.1.3. Asymmetric Dislocation Introduction

Asymmetric introduction of misfit dislocations has been observed in a number of III–V compounds[45,56] in that misfit dislocations are observed to form initially only along one ⟨110⟩ direction but not along the orthogonal one in (001) films. Abrahams *et al.*[57] first observed this effect in both $In_xGa_{1-x}P$ and $GaAs_{1-x}P_x$. A demonstration of this effect has already been shown in Figs. 14 and 15.

An interesting feature seen in Fig. 14 is the orthogonality of the initial misfit dislocations on each side of the tension–compression crossover. This effect is a direct consequence of the asymmetry of the zinc blende lattice and has the same explanation as asymmetric cracking.[10] Misfit dislocations on the tension side have Burger's vectors (or at least misfit-relieving components) opposite to those on the compression side. If the two were to lie along the same direction, then they would have to lie on opposite slip planes [i.e., (111) vs. $(\bar{1}\bar{1}\bar{1})$]. Since these planes each have different species of atoms, the two types of dislocations would possess different symmetries. However, if they were to lie along orthogonal directions, it can be shown[10] that they would be equivalent crystallographically, and thus presumably possess the same core energy,[58] which is a measure of the ease with which a dislocation can be introduced. Blanc[59] has considered this problem in detail and concludes that it is not yet possible to distinguish experimentally between the two cases, since precise knowledge of the core structure is currently unavailable. However, it is reasonable to conclude that since the core energies should be different, one type would requirc less energy for introduction than the other, and thus, the asymmetric dislocation distribution shown in Fig. 14 is consistent.

4.1.4. Misfit Dislocation Morphologies in Compositionally Graded Layers

In order to reduce the deleterious effects of lattice mismatch in III–V heterostructures such as $In_xGa_{1-x}P/GaP$, $In_xGa_{1-x}As/GaAs$, etc., compositional grading[60] has usually been employed. In this manner, the lattice constant (which in III–V compounds varies linearly with composition) is changed in either small abrupt steps or in a continuous manner from its initial value (which typically matches the substrate) to that of its final value. Both theoretical[1] and experimental[60,61] findings show that misfit dislocations will, in general, accompany this grading in order to relieve the lattice mismatch.

Figures 17 and 18 show two cross-sectional transmission-electron microscope (XTEM) micrographs[60] of VPE layers of $In_xGa_{1-x}P$ on a (100) GaP substrate which have been either step-graded (Fig. 17) or continuously graded (Fig. 18). The samples were prepared by inserting a preheated GaP substrate into the VPE deposition chamber, which had been set up to yield a layer of composition $\sim In_{0.05}Ga_{0.95}P$. In the case of step-graded layers, growth was continued for 5 min after which the HCl gas over the indium was abruptly increased to yield a crystal of composition $\sim In_{0.13}Ga_{0.87}P$. Thereafter, the In composition was increased in a similar manner at 5-min intervals until the final desired composition was reached. It is evident that the step-graded region consists of a series of narrow bands having high dislocation densities separated by wider regions of lower dislocation densities. The high-dislocation-density regions occur where the vapor composition was abruptly changed whereas the lower dislocation-density regions represent the ensuing growth after the abrupt step. It is clear that most of these dislocations are bent *normal* to the growth axis and thus do not propagate along the growth direction. Although inclined dislocations are present between the steps, they can almost always be seen to bend over into the interfacial plane at or before the next abrupt step.

The initial composition steps are relatively abrupt, whereas those of the later layers exhibit a noticeable lag in the compositional changes. Mass spectrometer measurements have shown[62] that as the HCl flow over either In or Ga in VPE reactors is increased to higher and higher values, the time required for complete equilibration of the metal and HCl becomes greater (from several seconds to several minutes). Thus, the lack of abruptness in the final steps is not surprising. The most important observation in this figure, however, is that the dislocation density in the constant composition region of the sample is below the limit of detection for TEM. This figure serves as dramatic proof that low-dislocation-density ($\leq 10^5/cm^2$) ternary alloys of arbitrary composition can be grown on a binary III–V substrate. Thus, if diffusion lengths were limited only by the spacing of dislocations, then diffusion lengths $\geq 10\ \mu m$ could, in principle, be achieved in these materials. The electrical properties in such crystals,

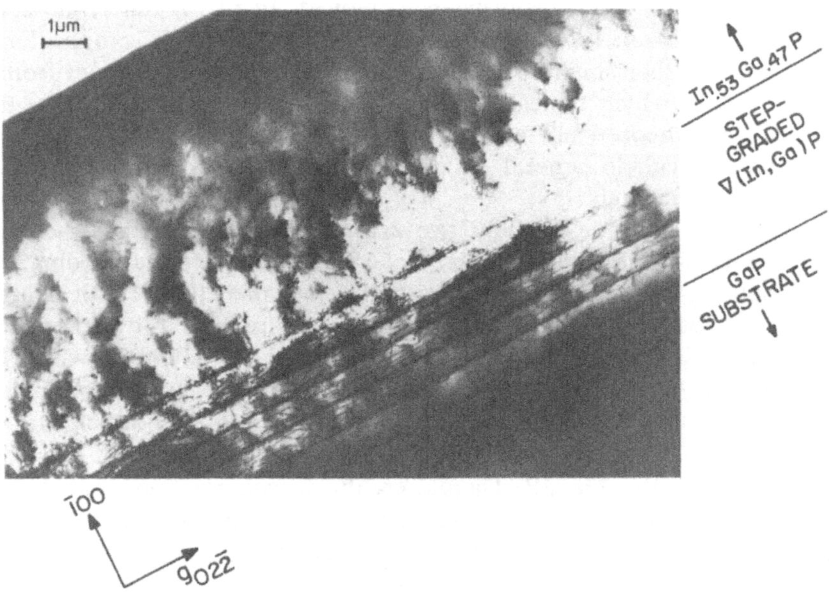

FIGURE 17. {011} XTEM micrograph of step-graded VPE $In_xGa_{1-x}P$ on a {100} GaP substrate. (Courtesy of the *Journal of Applied Physics.*)

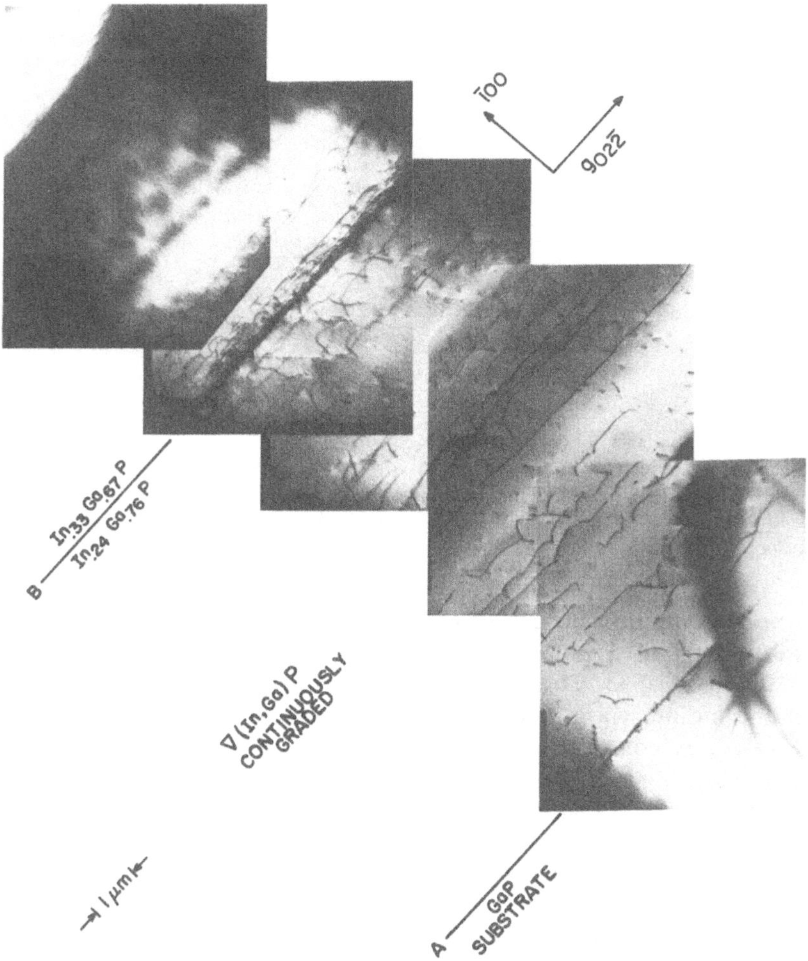

FIGURE 18. {011} XTEM montage of continuously graded VPE $In_xGa_{1-x}P$ on a {100} GaP substrate. (Courtesy of the *Journal of Applied Physics*.)

or in subsequently grown crystals, should then approach that of the bulk dislocation-free materials. Preliminary measurements of electron diffusion length in GaAs grown on structures such as that shown in Fig. 17 have indicated that this is indeed so.[63]

Although continuous grading often causes dislocations to propagate up through the constant composition region,[61] a final step-graded layer may be applied in order to bend these dislocations into the interfacial plane and thus eliminate them from subsequent layers. Figure 18 shows an electron micrograph of an $In_xGa_{1-x}P$/GaP layer in which the $In_xGa_{1-x}P$

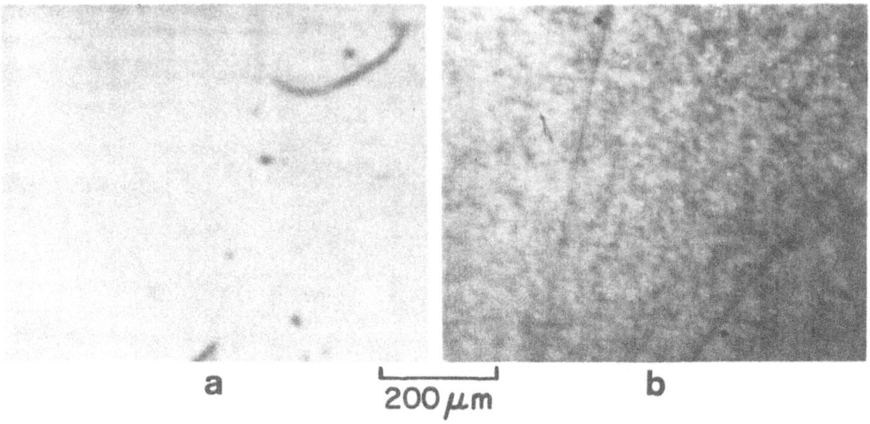

FIGURE 19. Reflection X-ray topographs of (a) the substrate and (b) the top of the epitaxial layer of a {100} VPE $In_{0.01}Ga_{0.99}As/GaAs$ composite. For the epi layer, $\rho_D \approx 0$, and for the substrate $\rho_D \approx 1 \times 10^4 \text{ cm}^{-2}$. The contrast features on the epitaxial layer correspond to surface defects (e.g., hillocks or scratches). (Courtesy of S. H. McFarlane.)

has been continuously graded (after an initial abrupt 5% In step at A) from $x-0.05$ to $x-0.24$, followed by another abrupt step at B from $x-0.24$ to $x-0.33$. The grade was achieved by changing the HCl flow over the indium via a motor-driven potentiometer on the mass flow control valve. A constant composition layer ($x-0.33$) was then deposited. The net grading rate was ~2% In per micrometer of growth. Using the calculation of Abrahams *et al.*,[61] a dislocation density $\sim 5 \times 10^8/\text{cm}^2$ is found in the graded region, which is equivalent to a dislocation spacing of ~0.5 μm. This is roughly the average spacing which can be measured in Fig. 18. However, once the abrupt compositional step has been made (B in Fig. 18), the dislocation density can again be seen to be $\leqslant 10^5/\text{cm}^2$ (a few dislocations extend about 1 μm above the step due to the previously mentioned inertia associated with the vapor-phase growth). Again, the outstanding feature of this figure is the paucity of dislocations in the constant composition region; the bending over of the dislocations away from the growth axis is again related to the abrupt compositional step.

Figure 19 provides direct experimental evidence that virtually dislocation-free epitaxial layers can be fabricated via the use of an abrupt lattice parameter step. This figure contains reflection X-ray topographs of both the top of the epitaxial layer and the underlying substrate of a 125-μm VPE deposit of $In_{0.01}Ga_{0.99}As$ on a (100) GaAs substrate. Although the substrate contained $\sim 1 \times 10^4 \text{ cm}^{-2}$ inclined 60° dislocations, all of these must have been bent over at the interface, due to the abrupt step ($\epsilon \sim 7 \times 10^{-4}$), since none are visible in the top layer (the few dark regions seen correspond to surface features such as scratches or hillocks). Thus, the dislocation density in the epitaxial layer is $<100\,cm^{-2}$ and is probably close to zero. It should be noted, however, that the composition of the epitaxial layer was uniform across the entire substrate. Any lateral compositional variations can give rise to inclined dislocations (see Fig. 22), and should be avoided if dislocations are to be completely removed.

Figure 20 depicts a (100) TEM micrograph of an ~2000-Å-thick ungraded layer of $In_{0.53}Ga_{0.47}P$ deposited on a GaAs substrate which was then selectively dissolved chemically in some areas. The orthogonal misfit dislocations are evident only where the substrate has not been dissolved, which provides further proof that dislocations in step-graded layers are confined to the interface and do not propagate into the epitaxial layer.

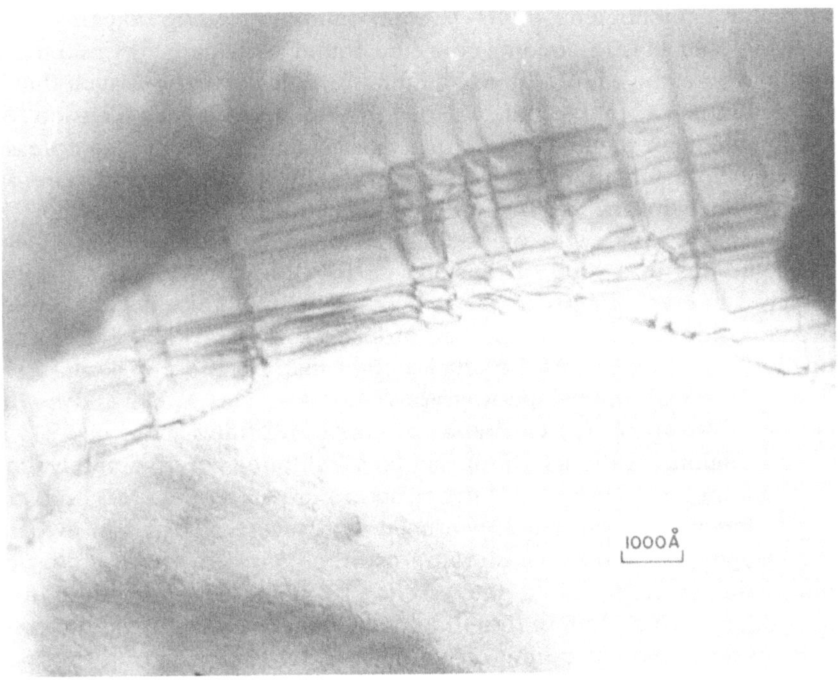

FIGURE 20. {100} TEM from a VPE $In_{0.53}Ga_{0.47}P$ layer on a {100} GaAs substrate. The bottom (clear) region corresponds to $In_{0.53}Ga_{0.47}P$, where the GaAs substrate had been completely removed chemically, and shows that misfit dislocations are combined to the interface and do not propagate into the epitaxial layer. (Courtesy of J. R. McGinn and M. D. Coutts.)

4.1.5. Tension versus Compression: Cracking

Most of the results in this chapter concerning VPE materials have been obtained from deposits of $GaAs_{1-x}P_x/GaP$, $In_xGa_{1-x}P/GaAs$ ($y > 0.5$), and $In_xGa_{1-x}As/GaAs$. However, these structures have one very important feature in common: all epitaxial layers are in a state of compression. Layers grown in tension (e.g., $GaAs_{1-x}P_x/GaAs$ and $In_xGa_{1-x}P/GaAs$, $y < 0.5$) have been found to crack asymmetrically whenever abrupt compositional steps have been introduced, and continuous grading at low grading rates has generally been employed to fabricate these structures. It is felt that the cracking phenomenon is a fundamental limitation which restricts the use of abrupt steps to epitaxial layers which are in compression as opposed to tension. Cracking in III–V compounds has been shown to be asymmetric. This asymmetry follows the asymmetry of the zinc blende structure, which lacks a fourfold symmetry axis in the (100) plane. A more detailed discussion of asymmetric phenomena in III–V compounds (including cracking, cross-hatching, and elastic bending) may be found elsewhere.[10] Cracking has been shown to be caused by the pileup of misfit dislocations such that the accumulation of extra half-planes forms a microcrack. Tension thus readily opens up the microcrack into a fracture line, whereas compression tends to close up the microcrack and thereby avoids fracture. The dark vertical lines in the XTEM micrograph in Fig. 2 correspond to dislocations associated with microcracks in the top $In_xGa_{1-x}P$ layer, since this layer was in a state of tension ($x \sim 0.46$). $In_xGa_{1-x}P/GaAs$ has been observed to crack readily for $x < 0.47$ ($\epsilon \geqslant 2 \times 10^{-3}$). This effect also relates to compositional grading, since, in order to avoid cracking, grading rates for tensile structures must be much longer than for those in compression. Figure 21 shows optical photographs[64] of two $GaAs_{0.6}P_{0.4}$ alloys which had been grown on (a) GaP and (b) GaAs substrates with only a 3-μm graded region. Although both samples exhibited approximately equal photoluminescent intensity,[40] the structure grown on the GaAs substrate clearly has poorer surface appearance with severe cracking in evidence. The rough appearance is due to disturbed growth effects,[10,65] which follow the cracking. Note the essentially unidirectional nature of the disturbances which follow the asymmetric cracking. The structure on the GaP substrate did not exhibit cracking. In order to grow a truly crack-free $GaAs_{0.6}P_{0.4}/GaAs$ structure, it is necessary to use an $\sim$30-μm graded region. Thus, a factor of 10 can be saved in growth time if compositionally graded III–V VPE alloys are grown in compression rather than under tension.

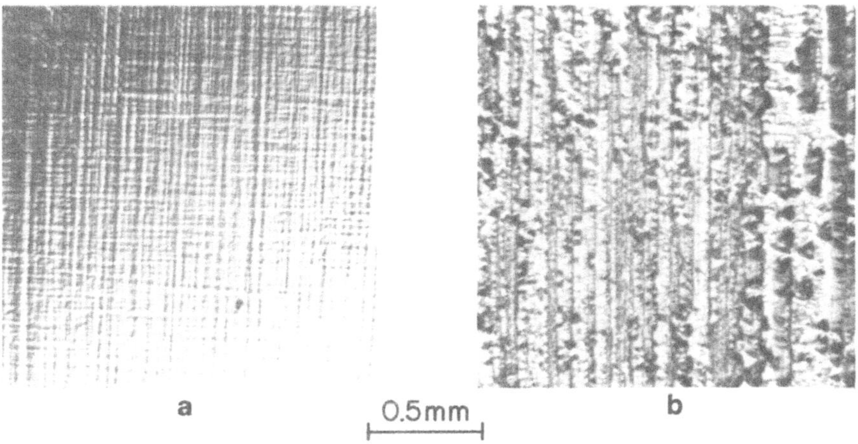

FIGURE 21. Optical photographs of two (100) VPE $GaAs_{0.65}P_{0.35}$ epitaxial layers that were compositionally graded over ~3 μm from a (a) GaP substrate and (b) GaAs substrate. The horizontal and vertical edges lie along ⟨011⟩ directions. (Courtesy of the *Journal of Crystal Growth.*)

It has also been found that mismatched layers in tension tend to have much higher densities of inclined dislocations than those in compression. This is most likely due to the propagation of the microcracks when they occur. Figure 22 shows a cathodoluminescence photograph of the two ends and middle of the $In_xGa_{1-x}P$/GaAs sample shown in Fig. 14 in which x varied from tension ($x \sim 0.46$) on one end to compression ($x \sim 0.53$) on the other. Since cathodoluminescence only reveals defects in direct bandgap materials that are within a diffusion length (~1 μm) of the surface, the sample was angle-lapped (1°) through the interface so that the substrate, interface, and top layer could all be examined. Notice first that the substrate dislocation density was not uniform but varied across the wafer from 2×10^4 to $16 \times 10^4\ cm^{-2}$. Note also the following:

1. The density of dislocations at the top of the compression region ($\epsilon \cong -3 \times 10^{-3}$) is a factor of 4 lower than the density in the substrate.
2. The dislocation density at the top of the "unstrained" region ($\epsilon \leqslant 1 \times 10^{-4}$) is within a factor of 2 of the substrate density.
3. The dislocation density at the top of the tension region ($\epsilon \simeq 2 \times 10^{-3}$) is three orders of magnitude higher than that in the substrate.

Although the number of dislocations in the compression region is clearly lower than that in the substrate, the fact that a reduction of only a factor of 4 was achieved might at first seem to contradict the results of the

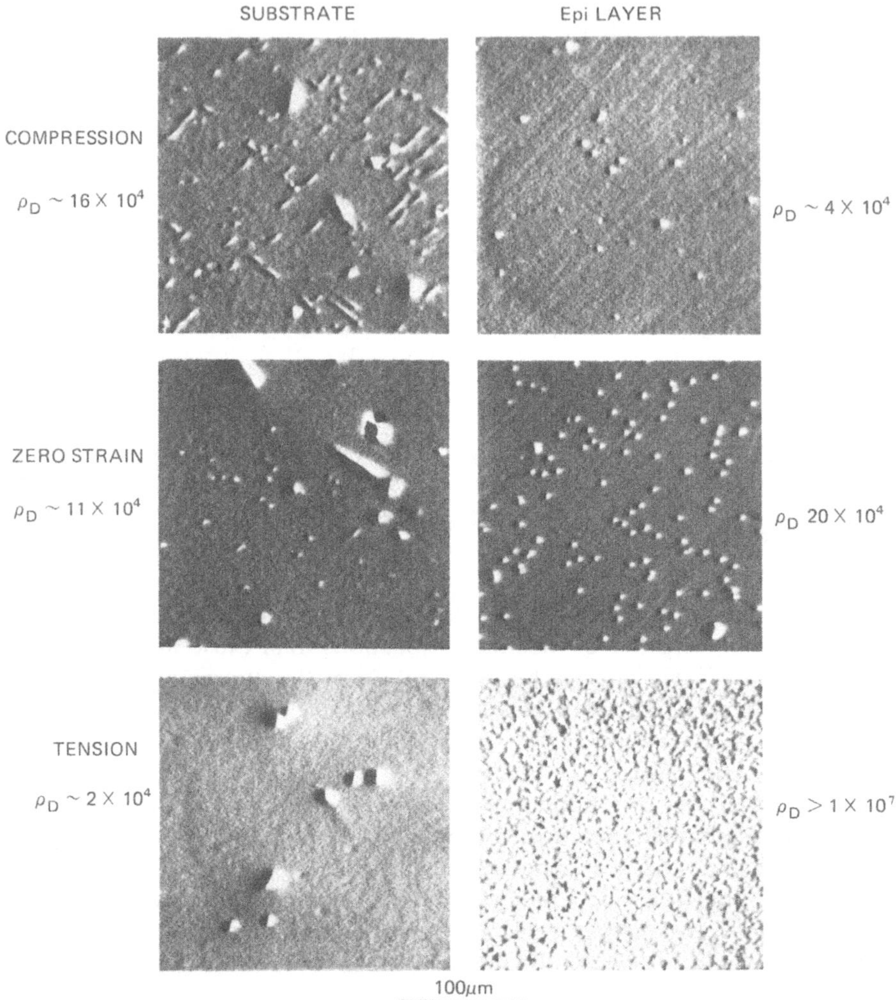

FIGURE 22. SEM cathodoluminescence photographs which show the inclined dislocation density (ρ_D) in both the substrate and top of the epitaxial layer in a {100} VPE $In_xGa_{1-x}P/GaAs$ sample in which x varied along the wafer as shown. (Courtesy of E. Levin.)

preceding section. However, since the lattice mismatch varied continuously across the wafer from compression to zero to tension, the density of interfacial misfit dislocations required would also vary across the sample. Thus, not all dislocations generated in a region of high mismatch would be required in an adjacent region of low mismatch, and the difference could be accommodated by the inclination of unnecessary dislocations away from the interface. The reasonable agreement of inclined dislocation density in the substrate and epitaxial layer of the "unstrained" region

($\epsilon \sim 0$) confirms the well-known effect[66] of dislocation propagation during crystal growth. The extra dislocations in the epilayer might well be misfit dislocations from the strained regions which have bent up. The orders of magnitude increase of dislocation density in the tension region is caused by dislocation generation and propagation via microcrack formation. This effect can also be seen in the XTEM micrograph of Fig. 2, which shows a high density of inclined dislocations in the tensile region.

4.1.6. Misorientation in Graded Layers

X-ray analysis has demonstrated a definite overgrowth/substrate misorientation in lattice-mismatched VPE layers.[44] The misorientation (α) varies from 0.5 to 2.0° and the axis of rotation lies either in the substrate plane or is somewhat inclined ($\approx$0–30°) to it. Often the rotation axis is not along a simple crystallographic direction and no correlation has yet been made between rotation axis and any crystal parameters. The effect has been noticed in VPE-grown lattice-mismatched deposits of $GaAs_{1-x}P_x$, $In_xGa_{1-x}P$, and $In_xGa_{1-x}As$. The misorientation was detected with a combination of X-ray methods which involve single-crystal diffractometry, Weissenberg, and Laue techniques. This misorientation should not be confused with the 3 to 5° misorientation that is introduced by mechanical polishing on most substrates used for VPE crystal growth.

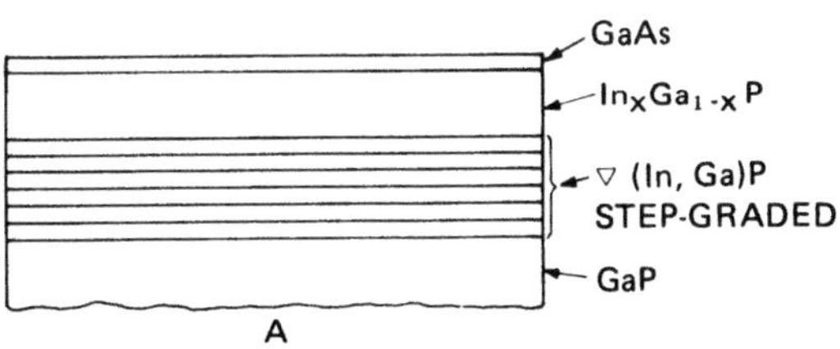

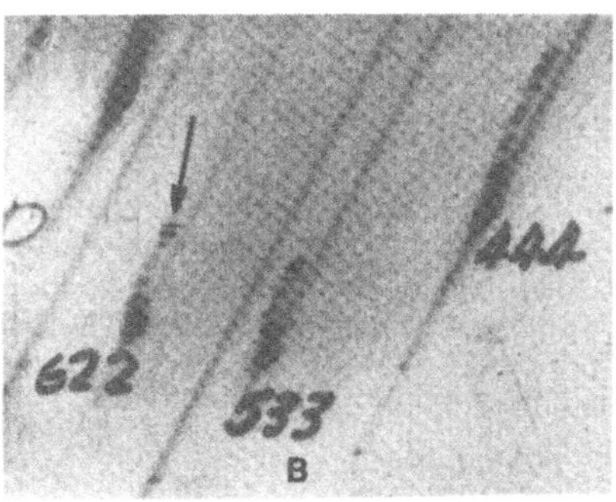

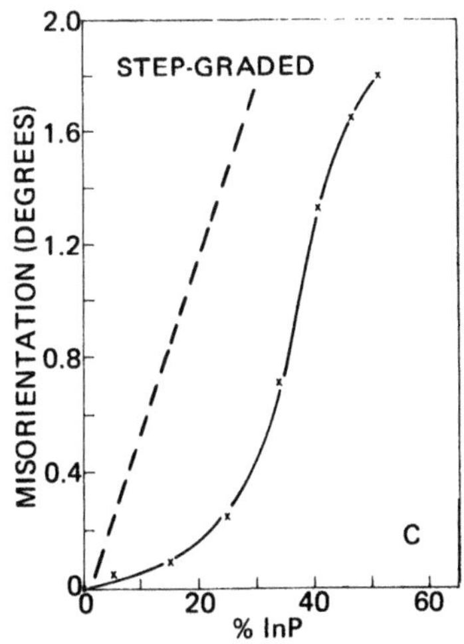

FIGURE 23. (A) Sketch, (B) Weissenberg photograph, and (C) a plot of misorientation (α) versus composition for a step-graded VPE layer of $In_xGa_{1-x}P$ on a {100} GaP substrate. (Courtesy of *Physica Status Solidi.*)

In the case of step-graded layers, X-ray diffraction has been obtained from each of the layers, thus permitting lattice constant and overgrowth/substrate misorientation to be determined. Figure 23 contains a sketch of the step-graded $In_xGa_{1-x}P$ layers, together with a Weissenberg photograph and a plot of misorientation (α) versus composition (x). The discrete nature of the misorientation–misfit relationship is apparent in this figure. Figure 24 contains similar information for a continuously graded $GaAs/In_xGa_{1-x}P/GaP$ sample whose uppermost alloy composition is nearly equal to that of the step-graded sample in Fig. 23. It is clear from Fig. 24 that both the lattice constant and the misorientation in continuously graded samples vary in a continuous, rather than discrete, manner. However, the magnitude of the misorientation is comparable to that of the step-graded layer. The solid lines in the misorientation versus

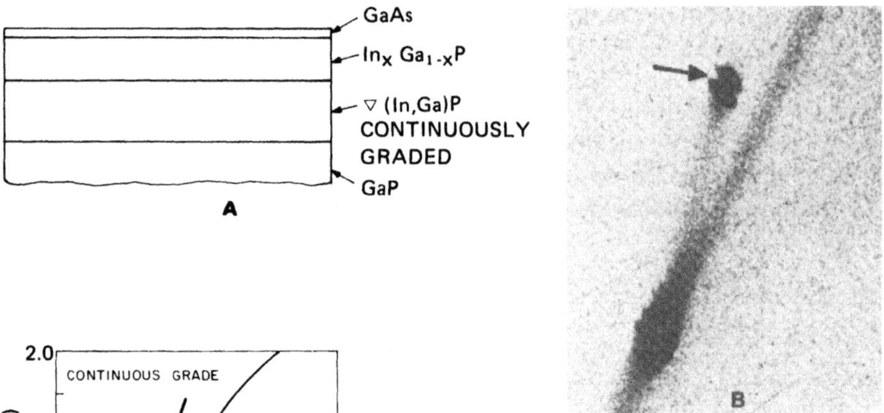

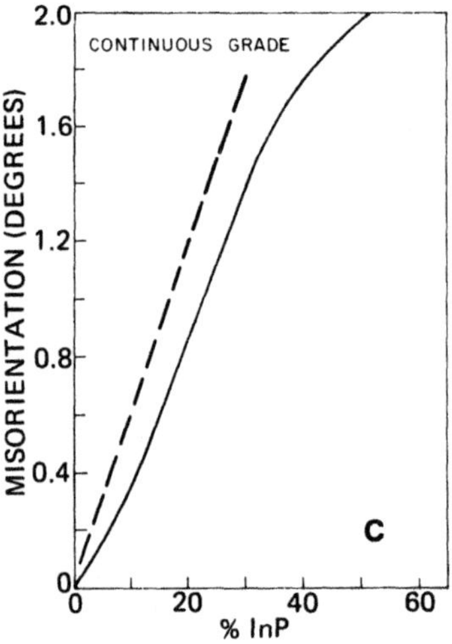

FIGURE 24. (A) Sketch, (B) Weissenberg photograph, and (C) a plot of misorientation (α) versus composition for a continuously graded VPE layer of $In_xGa_{1-x}P$ on a {100} GaP substrate. (Courtesy of *Physica Status Solidi.*)

composition plot represent the experimental results, and the dashed lines are calculated curves which are discussed later.

It is evident that the misorientation varies with compositional grading (i.e., lattice mismatch). This strongly suggests that the misorientation is related to the formation of misfit dislocations whose morphology has already been shown[60,61] to be related to the grading technique. Jesser[67] has recently performed an elegant vector analysis of closure failures associated with boundaries in crystals with differing lattice geometries. He suggests that in such crystals, three sets of dislocations may form at the boundary: one set (referred to as type 1) having Burger's vectors perpendicular to the boundary plane to accommodate tilt (forming the familiar tilt boundary), and another (type 2) with Burger's vectors lying in the interface to accommodate interfacial misfit. The third set—screw dislocations having Burger's vectors that lie in the interface and give rise to shear stresses and/or twist boundaries—is not considered here since no twist boundaries have been observed. Although a misfit-induced misorientation is not at all obvious, it can be shown that misorientation of the type just described must in general result from a parallel array of interfacial misfit dislocations. Consider such an array of dislocations having colinear but unspecified Burger's vectors (b). Each dislocation will, in general, have two edge components: one normal to the interface and one lying in the interface, as shown in Fig. 25. These edge components thus correspond to Jesser's type 1 and type 2 dislocations, only in this case both types are contained in one dislocation. The type 1 dislocation will give rise to a misorientation[68]

$$\alpha = b_1\rho_1 \tag{6}$$

where b_1 is the Burger's vector component perpendicular to the interface and ρ_1 is the linear dislocation density in the interface. Assuming that all misfit is relieved by dislocations, the linear dislocation density may in turn be related to the misfit strain (ϵ) from Eq. (2) as

$$1/s = \rho_1 = \epsilon/b_2 \tag{7}$$

where b_2 is the misfit-relieving Burger's vector component lying in the interface. Then,

$$\alpha = \epsilon b_1/b_2 \tag{8}$$

Equation (8) shows that the misorientation is directly proportional to the misfit strain. Thus, misorientation normal to the interface will in general occur, except for the special case when misfit is relieved by pure edge dislocations whose Burger's vectors lie entirely within the interface.

In III–V compounds, misfit relief has been shown[61,65] to be

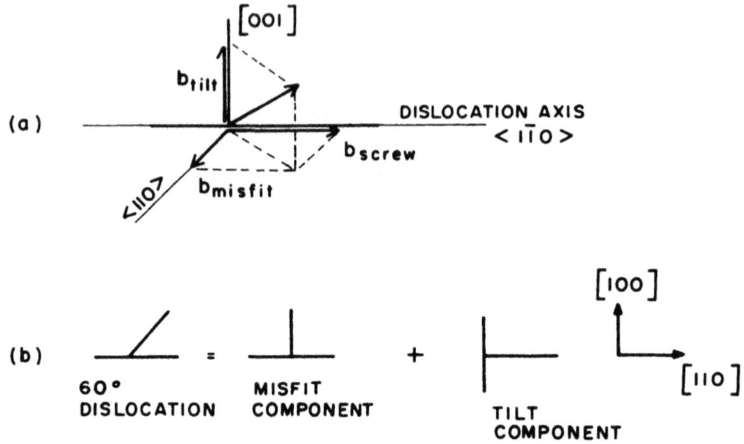

FIGURE 25. (a) The three-dimensional Burger's vector components of a 60° dislocation. (b) A view along the dislocation axis which shows the breakup of the Burger's vector into misfit and tilt accommodating components. (Courtesy of the *Journal of Crystal Growth.*)

accomplished largely by 60°-type dislocations that have both type 1 and type 2 Burger's vector components. For the case of a step-graded layer of (100) $In_{0.5}Ga_{0.5}P/GaP$, the total misfit strain is ≈ 0.037. Assuming an interfacial dislocation along an [011] direction having $b = [101]$, the tilt component ($|b_1|$) along [001] would be $|b|/2^{1/2}$. Similarly, $|b_2|$ would be $|b|/2$. Thus,

$$\alpha = 1.41\epsilon = 0.052 \text{ rad} \approx 3° \tag{9}$$

This, of course, represents the maximum value of misorientation, since dislocations with colinear Burger's vectors were assumed in Eq. (8). (The misorientation will be distributed over the graded region, as will the misfit dislocations, in accordance with the grading technique used to grow the crystal.) Since in real crystals, not all misfit dislocations are either 60° type nor all of like sign, the observed variation of 0.5 to 2.0° misorientation is not surprising. Also, since the distribution of misfit dislocations is never completely uniform, and is in fact asymmetric in III–V compounds (see Figs. 14 and 20), the observed variation in axis of rotation is also to be expected. The dashed lines in Figs. 23 and 24 represent $\alpha = 1.41\epsilon$, which is the predicted misorientation for an equilibrium distribution of 60° misfit dislocations. The continuous grade evidently approached this situation more closely than the step grade. However, it can be seen in Figs. 17 and 18 that misfit dislocations are more uniformly distributed in continuously graded layers than in step-graded layers, and therefore would probably relieve strain more uniformly and completely. The treatment is also valid for inclined dislocations as long as appropriate Burger's vector components are considered. However, it is clear that misorientation normal to the interface is a general phenomenon associated with compositional grading in III–V compounds prepared via VPE.

4.2. Dislocation Morphology in LPE Materials

As one might expect from the great difference in LPE and VPE growth techniques, there is generally a difference in the morphology of misfit dislocations in heteroepitaxial layers prepared by these two methods. In some LPE layers (e.g., $Al_xGa_{1-x}As_yP_{1-y}$ on GaAs,[45] $InAs_{1-x}P_x$ on InP,[69] and $In_xGa_{1-x}As_yP_{1-y}$ on InP[22]), the misfit is relieved by a crosshatched array of dislocations similar to that discussed in detail in Section 4.1 for VPE material. More generally, however, the misfit dislocations in lattice-mismatched LPE materials align to form a subgrain structure within the essentially single-crystal epitaxial layer.

The difference in dislocation morphology between lattice-mismatched LPE and VPE heteroepitaxial layers is most vividly demonstrated for $In_xGa_{1-x}As$ layers grown on GaAs substrates.[71] Illustrated in Fig. 26 are two reflection X-ray topographs of $In_{0.06}Ga_{0.94}As$ layers grown on GaAs by standard LPE and VPE techniques. The room temperature lattice mismatch between substrate and epitaxial layer is 4×10^{-3}. The

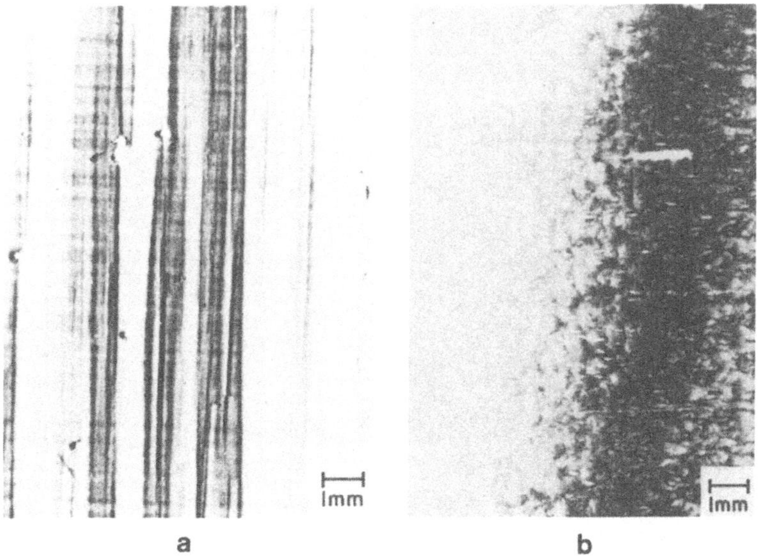

FIGURE 26. Reflection X-ray topograph of $In_{0.06}Ga_{0.94}As$ grown on {100} GaAs by (a) VPE and (b) LPE techniques. (Courtesy of the *Journal of Electronic Materials*.)

topograph of the VPE layer shows, as discussed in the preceding section, the orthogonal array of strain-relieving dislocations. The LPE-grown sample shows a dislocation array that forms a subgrain structure. The subgrains are about 10 μm in diameter and are misoriented from one another by $\leq 0.2°$. However, no overall substrate overgrowth misorientation has been observed in LPE layers. For $In_xGa_{1-x}As$ alloys, the dislocation morphology illustrated in Fig. 26 for the two growth techniques is representative of alloy compositions from $x = 0.03$ to 0.30.

Another important point illustrated in the $In_xGa_{1-x}As$ system is the difference in the effect of step-grading by the two techniques. Step-grading in VPE materials seems to be much more effective in reducing the dislocation density in the topmost active regions. In Fig. 27 are

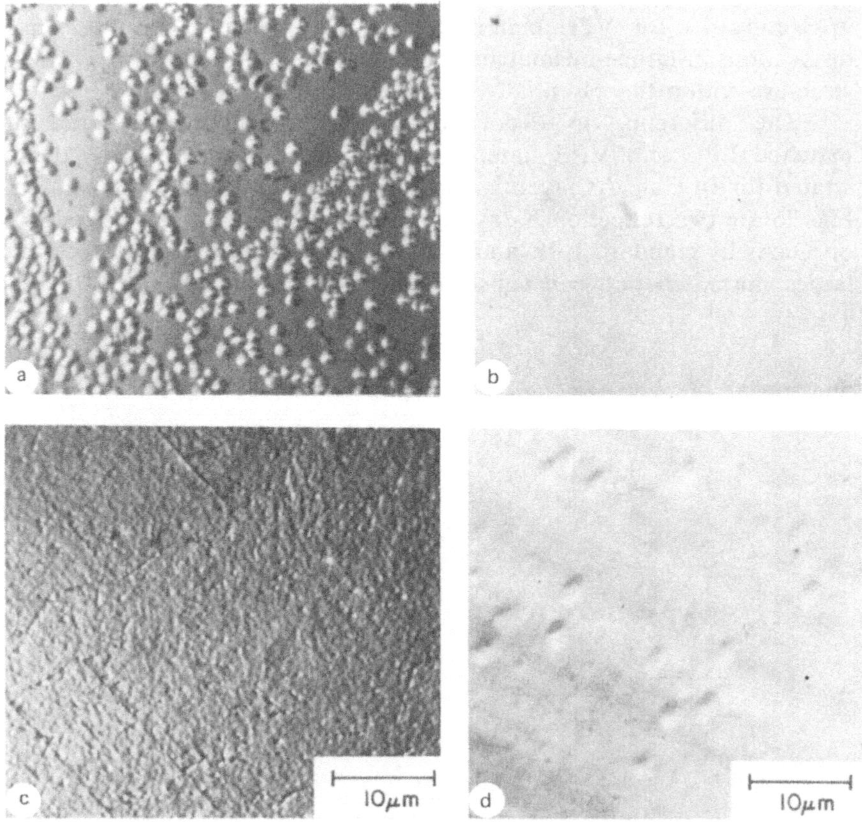

FIGURE 27. Optical photograph of etch pits on the surface of LPE (left) and VPE (right) $In_xGa_{1-x}As$ grown by step-grading on {100} GaAs substrates. (a) $In_{0.06}Ga_{0.94}As$—LPE; (b) $In_{0.06}Ga_{0.94}As$—VPE; (c) $In_{0.24}Ga_{0.76}As$—LPE; (d) $In_{0.28}Ga_{0.72}As$—VPE. (Courtesy of the *Journal of Electronic Materials.*)

Table 4. Comparison of LPE- and VPE-Grown Layers of $In_xGa_{1-x}As$

Growth method and top layer composition (% InAs)		Intermediate grading layers (% InAs)	Dislocation density in top layer (cm^{-2})	Average dislocation spacing[a] (μm)	Measured diffusion length (μm)
VPE	~3	None	3×10^4	57	—
	6.1	None	5×10^4	44	2.6
	14.0	4.7, 6.7	5×10^6	4.5	2.3
	17.7	—	2×10^5	22.3	—
	20.5	4.7, 6.7, 9.7, 12.5, 14.7	1.5×10^7	2.5	2.7
	23	—	3×10^5	18.2	—
	28.4	2.4, 4.8, 8.5, 11.5, 14.5, 20	2×10^6	7.1	2.6
LPE	6.6	None	1.5×10^7	2.5	—
	8.1	None	5×10^6	4.5	—
	13.8	6.3	6×10^7	1.3	—
	23.8	5.7, 13.7	$>10^8$	<1	0.5

[a]Taken as the square root of the reciprocal of the dislocation density. Dashes indicate parameters that were not measured.

photomicrographs of the surfaces of LPE- and VPE-grown $In_xGa_{1-x}As$ layers after etching in A–B dislocation etch.[72] As can be clearly seen for comparable compositions and thus lattice misfit, the VPE technique produces a significantly lower defect density. Even lower densities are possible in VPE layers if uniform composition is maintained across the entire wafer. The comparison between VPE and LPE is shown in more detail in Table 4, where detailed data on the dislocation density by etch pit count and composition of the steps are given. All of the growths were made on GaAs substrates with dislocation densities $<3\times10^3\,cm^{-2}$. The average dislocation spacing is also included in Table 4. Since dislocations act as nonradiative recombination centers, if their spacing is greater than the minority carrier diffusion length, which is usually a few micrometers in most III–V materials, then the dislocations will have little effect on the properties of optoelectronic devices made from these materials. However, in terms of device reliability, even one dislocation may be important (i.e., the gradual degradation of GaAs cw lasers by dark-line damage[73]), and the dislocation spacing and device area fix the yield of long-lived devices.

In the comparison of VPE and LPE $In_xGa_{1-x}As$, the GaAs face orientation which produced the most consistently high-quality layers were used [3° from the (100) toward the ⟨110⟩ direction for the VPE layers and the (111)A for the LPE layers]. This difference in orientation does not seem to be a key factor. LPE growth on the (111)B surface produced comparable dislocation densities to that discussed here for the (111)A. However, higher luminescent efficiencies have been reported(74) for $In_xGa_{1-x}As$ LEDs on the (111)B, although dislocation densities were not reported for these structures.

Although step-grading does not appear to be as effective in LPE growth as in VPE growth, simple growth interruptions without changing the composition in LPE growth can be used to reduce dislocation densities. The growth interruptions can be achieved by stopping and restarting the cooling(75) or simply by moving the sample from bin to bin in a multibin slider boat.(76) In homoepitaxial GaP growth, dislocation densities were reduced from $\sim 5 \times 10^5$ to $\sim 1 \times 10^5\,cm^{-2}$ by such cooling rate interruptions. Similarly, in the heteroepitaxial growth of GaAs on GaP, densities were reduced from 6×10^7 to $5 \times 10^6\,cm^{-2}$ by multibin epitaxy. This technique is effective, providing about a factor of 2 for each interruption, but does not seem capable of reducing dislocation densities much lower than $10^5\,cm^{-2}$. The LPE growth interruptions bend dislocations over at the interface, but are not as effective as VPE step-grading. The origin of the dislocation reduction in LPE-grown material has not been extensively investigated and thus is not yet well understood.

The subgrain structure of LPE heteroepitaxial layers arises either from a somewhat distorted nucleation process (because of the lattice mismatch) or from polygonization (the movement and alignment of dislocations) during growth. Evidence for both mechanisms seems to exist from the growth of $Al_xGa_{1-x}As$ on GaP. Figure 28 shows an {011} XTEM micrograph of an LPE layer of $Al_xGa_{1-x}As$ on a (111) GaP substrate.(64) Note the high density ($\sim 10^{10}\,cm^{-2}$) of inclined dislocations which propagate away from the interface along the growth direction. This density can be reduced substantially by growing additional $Al_xGa_{1-x}As$ layers onto the original one. Figure 29 contains a {111} TEM micrograph of the GaAs layer of a (111) LPE GaAs/$Al_xGa_{1-x}As$/GaP structure in which three successive $Al_xGa_{1-x}As$ layers had been deposited. The dislocation density in this layer is considerably lower than that at the $Al_xGa_{1-x}As$/GaP interface. Dislocations are now seen to align in low-energy configurations of low-angle (~0.1–0.2°) grain boundaries. A misorientation of ~0.1° was measured across the boundary via the Kikuchi line shift(77) seen in the insets to Fig. 29. This value is in reasonable agreement with that calculated from the average dislocation spacing and also with values deduced from X-ray topographs such as the one seen in Fig. 30. It was shown by van Oirschot *et al.*(78) that for roughened GaP substrates, the subgrain size is directly related to the surface roughness, indicative of a nucleation phenomenon. However, a

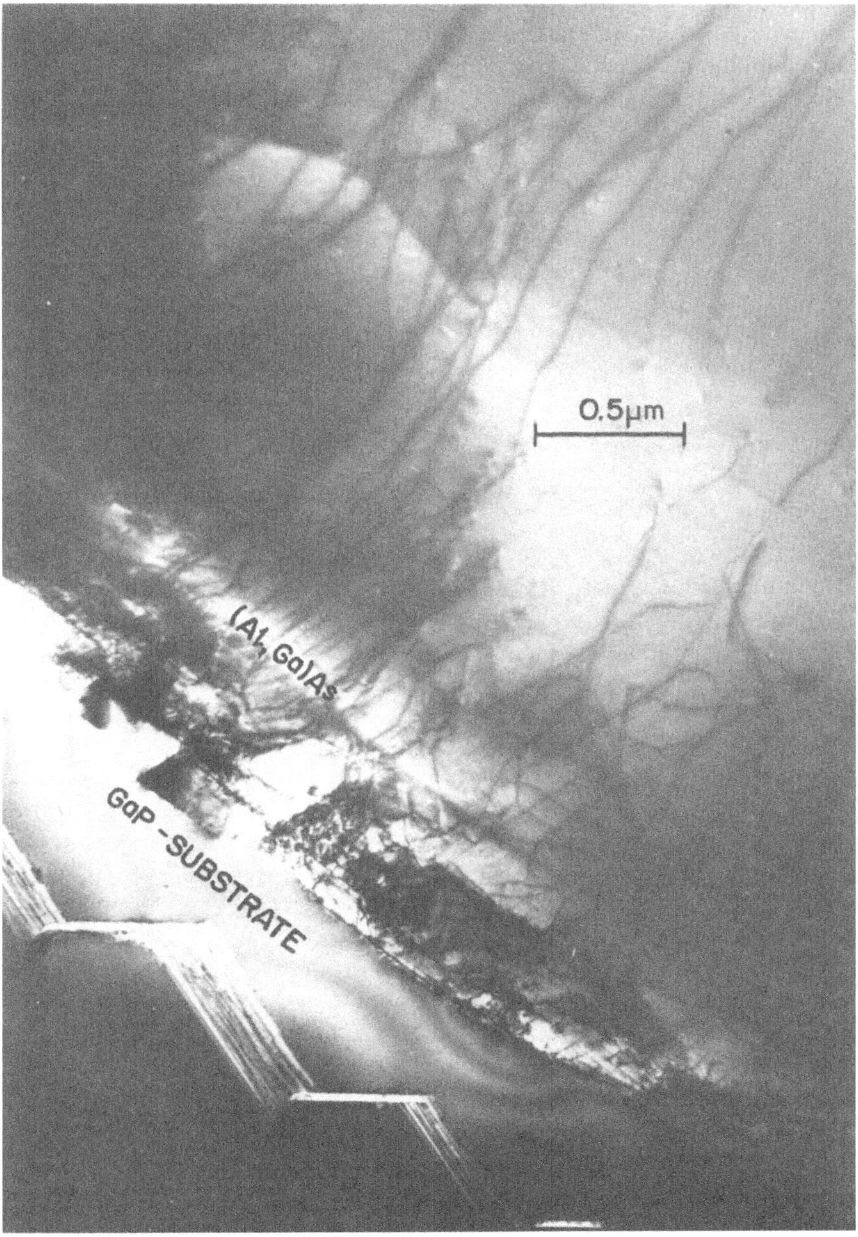

FIGURE 28. {011} XTEM micrograph of (AlGa)As deposited directly on a {111} GaP wafer via LPE. (Courtesy of the *Journal of Crystal Growth.*)

subgrain structure has also been observed for the growth of GaAs on GaAs[63] if the GaAs substrate surface is roughened enough to increase substantially the dislocation density in the subsequent GaAs epitaxial layer. This result is indicative of a polygonization mechanism, which should only occur at high dislocation densities.

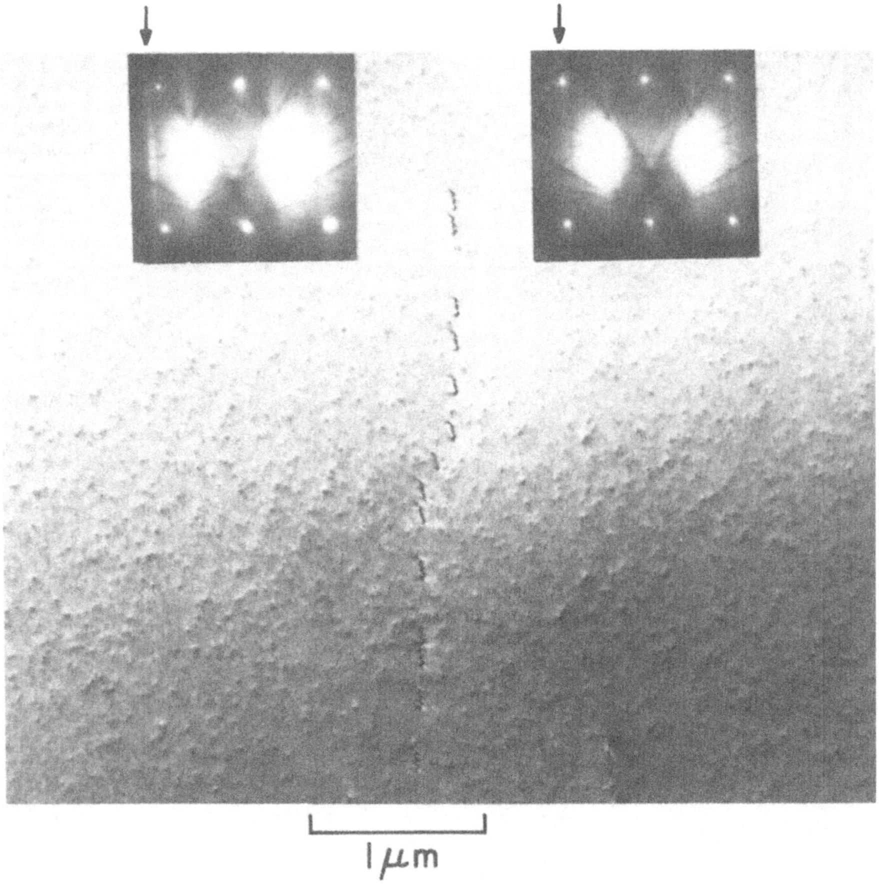

FIGURE 29. {111} TEM micrograph from the GaAs region of a {111} LPE GaAs/$Al_xGa_{1-x}As$/GaP multilayered structure. The dislocations have aligned into a low-energy configuration and the misorientation (~0.1°) across the boundary can be seen via the Kikuchi line shift indicated by the arrows on the inset. (Courtesy of M. S. Abrahams and C. J. Buiocchi.)

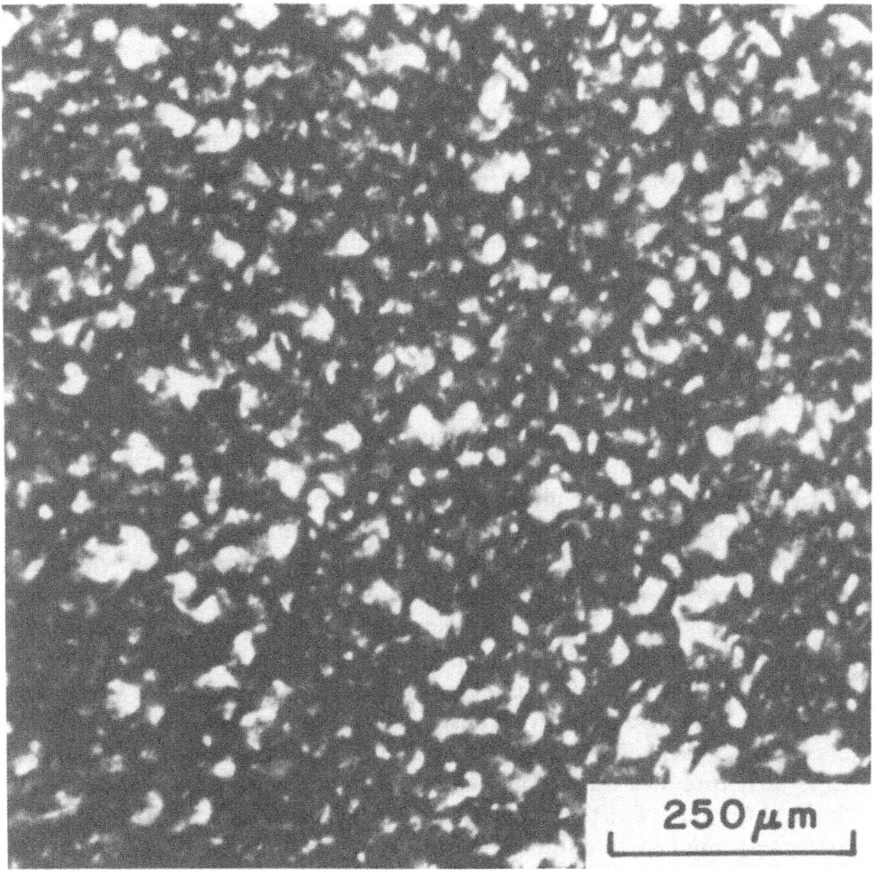

FIGURE 30. Reflection X-ray topograph of a (111) GaAs/$Al_xGa_{1-x}As$/GaP LPE sample showing many subgrains which are misoriented ~0.1–0.2°. (Courtesy of the *Journal of Crystal Growth*.)

5. APPLICATION TO DEVICE STRUCTURES: TRANSMISSION PHOTOCATHODES

Most of the microstructural studies described in this chapter were motivated by the desire to optimize crystal growth techniques in order to improve electrooptical properties of device structures. One example has already been shown in Fig. 2 (which is an XTEM micrograph of an actual VPE double-heterojunction laser), whereby the deleterious effects of preheating GaAs in pure hydrogen (rather than arsine) can readily be seen. Another useful result learned from this figure is that almost all of the dislocations in the active GaAs layer originate from the cracks in the upper N–$In_xGa_{1-x}P$ layer (none seem to originate at or propagate from the lattice-matched P–$In_{0.5}Ga_{0.5}P$/GaAs interface). Thus, in multilayered heterostructures, dislocations in one layer can originate in subsequently grown layers as well as in previously grown ones and lattice-matched interfaces should always be used for minimum dislocation content.

Many of the TEM micrographs and X-ray topographs in this chapter were obtained from GaAs transmission photocathodes. This structure[79–82] consists of a thin GaAs layer ($\sim 1\,\mu$m thick) deposited on a light transparent ($\lambda > 7000$ Å) substrate. In operation, light impinges onto the substrate and is absorbed in the GaAs, creating photoelectrons, which are extracted from the free GaAs surface that has been treated with Cs and O_2 to create a negative electron affinity (NEA) surface.[83] For this device to be effective, it must allow efficient transport of electrons from where they are created near the substrate–cathode interface to the free NEA surface. This requires a low value of interfacial recombination velocity (s_b) at the interface and a long diffusion length (L) for electrons in the GaAs. Since there is no convenient transparent substrate which lattice-matches GaAs, the cathode is often grown on non-lattice-matching substrates such as GaP. To reduce misfit effects, a buffer layer is grown between the substrate and GaAs layer. These devices are prepared via LPE[79] as GaAs/$Al_xGa_{1-x}As$/GaP and via VPE[81] as GaAs/$In_xGa_{1-x}P$/GaP. The LPE structure has almost all the 3.7% lattice mismatch concentrated at the $Al_xGa_{1-x}As$/GaP interface. Although the dislocation density is extremely high at this interface, the use of multiple $Al_xGa_{1-x}As$ reduces this density below the level at which L is greatly affected[76] (i.e., $<10^7\ cm^{-2}$; see Figs. 28 and 29). At the active $Al_xGa_{1-x}As$/GaAs interface, the excellent lattice match between $Al_xGa_{1-x}As$/GaAs provides a low interfacial recombination velocity. In the VPE-grown structure, the 3.7% misfit dislocations are confined to the buffer layer by grading the $In_xGa_{1-x}P$ either stepwise or continuously (see Figs. 17 and 18). The structure is sketched in Figs. 23 and 24. Two critical features must be present in the VPE-grown cathodes. The misfit dislocations must be effectively confined to the graded layer and the $In_xGa_{1-x}P$/GaAs must be well lattice-matched. This latter requirement calls for precise compositional control in the $In_xGa_{1-x}P$, whereas $Al_xGa_{1-x}As$/GaAs has no such requirement. Transmission photocathodes can be evaluated by considering the quantum yield in the transmission and reflection mode. This information allows L and s_b to be evaluated from yield curve analysis.[84]

The ratio of transmission (S_T) to reflection (S_R) sensitivity, diffusion length, and interfacial recombination velocity for a group of VPE $In_xGa_{1-x}P/GaAs$ photocathodes are plotted versus x in Fig. 31. Clearly, the ratio S_T/S_R, and L and s_b are all measures of crystal perfection. All of these parameters decrease from their optimum values around a critical $In_xGa_{1-x}P$ composition ($x = 0.53$). This composition is not exactly at the lattice match because, as discussed in Section 2.2.1, there is a small amount of In which diffuses into the GaAs so that the actual $In_xGa_{1-x}P$ lattice-matched composition would be somewhat higher in indium. Figure 4 was obtained from a lattice-mismatched transmission photocathode and shows how dislocations are generated at the interface even though the $In_xGa_{1-x}P$ constant composition region was dislocation-free.

These results dramatically illustrate the sensitivity of device performance to lattice mismatch and resulting crystal imperfections. Compositional variation of as little as 2% (ϵ as small as 7×10^{-4}) results in a dramatic decrease in sensitivity ratio by decreasing the diffusion length and increasing interface recombination.

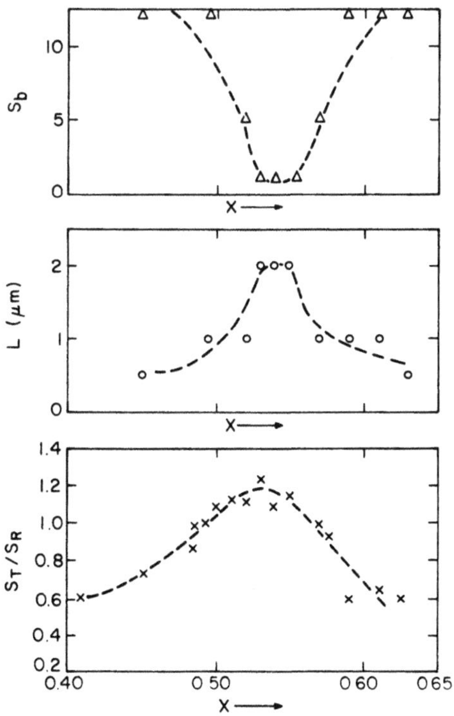

FIGURE 31. Interfacial recombination velocity (s_b), diffusion length (L), and transmission to reflection sensitivity ratio (S_T/S_R) vs. composition (x) for $In_xGa_{1-x}P/GaAs$ transmission photocathodes.

6. SUMMARY

The use of compositional grading techniques to reduce dislocation densities is one of the most important aspects of this chapter and should have application in all heteroepitaxial systems. The principles learned from III–V compounds could easily be extended to other alloy systems. The VPE growth of Ge_xSi_{1-x} alloys would be just one example. Silicon might in general be chosen as the substrate, since it has the smaller lattice parameter and would therefore place the epitaxial layer in compression (to avoid cracking). Mixed hydrides of silicon and germanium could then be used to step-grade up to some desired alloy composition, with dislocations being eliminated or minimized in the ensuing layers. The technique is also applicable to metal–metal systems.

The density and morphology of dislocations in comparable VPE and LPE III–V compounds are different and, as the technology stands today, the most effective control of dislocations is with vapor-phase epitaxy. This advantage probably arises from the almost unlimited control of the composition grading during vapor-phase epitaxy. Both technologies, however, are capable of producing heteroepitaxial layers in systems with dislocation densities low enough so that there is no first-order effect on most optoelectronic devices. In fact, VPE-grown $In_xGa_{1-x}P/GaAs$[(85)] and $In_xGa_{1-x}P/In_xGa_{1-x}As$[(86)] double-heterojunction room temperature cw lasers have been produced with properties comparable to LPE $Al_xGa_{1-x}As/GaAs$ lasers, while LPE has been used to produce $InP/In_xGa_{1-x}As_yP_{1-y}$[(87)] and $Al_xGa_{1-x}As_ySb_{1-y}/GaAs_ySb_{1-y}$[(88)] double-heterojunction cw lasers. But, if long-lived cw lasers are to be produced in these systems, the dislocation density will still probably have to be further reduced to a point at which the completed devices have only a small probability of containing a single dislocation.

The study of elastic strain in heteroepitaxial III–V compounds has also produced some important results. The calculation of stresses in multilayered structures should prove useful even in noncrystalline systems as long as the strains between layers can be specified. Tetragonal distortion and its related bandgap shifts are present in virtually all heteroepitaxial systems, although their magnitudes are often small. Possible device applications have yet to be exploited.

This chapter has demonstrated that both the elastic strain and dislocation content of III–V heterostructures (and their relevant manifestations) depend strongly on crystal growth parameters such as growth temperature and compositional grading. However, the findings described here should in no way be viewed as limited to III–V compounds alone.

Indeed, most of the results are of a general nature and apply to any crystalline system.

ACKNOWLEDGMENTS

The authors would like to acknowledge the assistance of many people at RCA Laboratories who made this chapter possible. Special acknowledgments are due to M. S. Abrahams and C. J. Buiocchi for electron microscopy; T. J. Zamerowski and L. Elbaum for crystal growth; S. H. McFarlane and R. T. Smith for X-ray work; E. Levin for cathodoluminescence; C. J. Nuese and J. Blanc for criticism of the manuscript; and C. A. Rossell for preparation of the manuscript.

APPENDIX A. STRESSES IN HETEROEPITAXIAL LAYERS

A heteroepitaxial structure composed of layers having different material properties (such as lattice constant or thermal expansion coefficient) will in general contain elastic strain. This strain in turn gives rise to stresses within each layer and to an overall bending of the composite itself. The following calculation(26) attempts to evaluate the stress at any position in any layer, and to predict the radius of curvature of the composite.

Consider a number of heteroepitaxial layers joined together as shown in Fig. 6. If the unconstrained length of each of the layers is different and the layers are joined and constrained to match at the interfaces, the composite will bend and assume some radius of curvature R and longitudinal stresses σ_i will be set up in each layer.(89) The strains ϵ_{ij} might be due to a difference in lattice constant, thermal expansion coefficient, or any other mechanism which gives rise to elastic strain. However, as long as the strains can be specified and elastic behavior assumed, both R and σ_i can, in principle, be calculated. The calculation for a long, narrow N-layer composite, such as that shown in Fig. 6, will be described since this structure is of frequent practical interest.

The structure shown in Fig. 6 is analogous to the "bimetallic strip" first treated by Timoshenko.(25) Saul(24) has calculated R and σ_i for the uppermost layer of a GaAs–GaP composite structure of similar shape which was strained due to thermal expansion differences. The present calculation will basically follow the method of Saul generalized to include elastic strains of arbitrary nature and can be used to calculate the stress at any point in the layer.

In the calculation of bending stresses, it is commonly assumed that each layer is composed of an infinite number of elastic "fibers" that lie along the length of the composite (i.e., parallel to the top and bottom surfaces) and that the strain in each fiber is the same all along the length of the layer.[89] Since the force F_i in each layer is caused by the action of external members (e.g., other layers), it may be assumed constant throughout the layer. (If it were not constant, the number of atomic planes at the top and bottom would differ, and dislocations would be required *within* the layer.) These assumptions are reasonable for the case of crystalline layers, since the applied strains ϵ_{ij} act in the interfacial plane which is parallel to the fibers. Furthermore, since $L \gg w$, bending along the width (w) may be neglected so that the problem reduces to one dimension. (Since only Poisson strains exist normal to the layers, no stresses exist along this direction in any case.) The composite of Fig. 6 may thus be described as having forces F_i and bending moments M_i in each layer as shown. R, F_i, and M_i are all drawn in the positive sense. The actual signs of each are part of the solution. For the case of a three-layer structure, mechanical equilibrium demands

$$\Sigma F = 0: \quad F_1 + F_2 + F_3 = 0 \tag{A.1}$$

$$\Sigma M = 0: \quad M_1 + M_2 + M_3 + F_1(t_1/2) + F_2(t_1 + t_2/2) + F_3(t_1 + t_2 + t_3/2) = 0 \tag{A.2}$$

where the moment due to each force F_i is assumed to act through the neutral axis (middle) of each layer (see later).

The bending moment M_i may be shown[89] to be

$$M_i = E_i I_i / R = E_i t_i^3 w / 12R \tag{A.3}$$

where E_i is Young's modulus, I_i is the moment of inertia of an area $t_i w$, and t_i is the thickness of the ith layer. The boundary conditions at each interface which generate the final equations necessary for the solution are worth some explanation. Consider two layers of different materials, each bent to a radius R (Fig. A1). Elasticity theory predicts[89] that (a) the neutral axis (N.A.) of a given layer lies in the middle of the composite and defines the fiber whose length remains unchanged due to bending; and (b) the strain *due to bending alone* is given as $\epsilon_B = x/R$, where x is the distance away from the neutral axis. Thus, the strain at the top of each layer is given as

$$\epsilon_{\text{top}} = +t_i/2R \qquad \text{(tensile)} \tag{A.4}$$

whereas that at the bottom is $\epsilon_{\text{bottom}} = -t_i/2R$ (compressive).

If the two layers are then joined together such that atom-by-atom alignment is maintained, and the composite is kept at the same radius R, the internal forces F_i and F_j then adjust to maintain coherency. The total

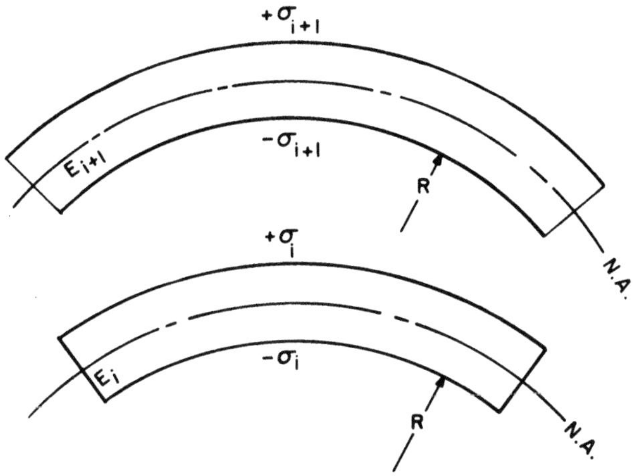

FIGURE A1. Stress distribution in two dissimilar strips bent to the same radius of curvature *R*.

strain in the uppermost fiber in the *i*th layer is then[89]

$$\epsilon_i = \frac{F_i}{Ewt_i} + \frac{t_i}{2R} \qquad \text{(A.5)}$$

Similarly, the total strain in the bottom fiber of the $(i+1)$th layer is

$$\epsilon_{i+1} = \frac{F_{i+1}}{E_{i+1}wt_{i+1}} - \frac{t_{i+1}}{2R} \qquad \text{(A.6)}$$

The difference

$$\epsilon_i - \epsilon_{i+1} \qquad \text{(A.7)}$$

reflects the difference in lattice spacing between the two layers. Thus

$$\epsilon_i - \epsilon_{i+1} = \frac{F_{i+1}}{E_{i+1}wt_{i+1}} - \frac{F_i}{E_iwt_i} - \frac{1}{2R}(t_i + t_{i+1}) \qquad \text{(A.8)}$$

In an N-layer composite, this gives rise to $N-1$ equations:

$$\sum_{i=1}^{N-1}\left[\epsilon_i - \epsilon_{i+1} = \frac{F_{i+1}}{E_{i+1}wt_{i+1}} - \frac{F_i}{E_iwt_i} - \frac{1}{2R}(t_i + t_{i+1})\right] \qquad \text{(A.9)}$$

These can be coupled with Eqs. (A.1) and (A.2) so that the $N+1$ variables (N stresses $+ R$) may be solved:

$$\sum_{i=1}^{N} F_i = 0 \qquad \text{(A.10)}$$

$$\sum_{i=1}^{N}\left[\frac{E_i t_i^3 w}{12R} + F_i \sum_{j=1}^{i}\left(t_j - \frac{t_i}{2}\right)\right] = 0 \tag{A.11}$$

$$\sum_{i=1}^{N-1}\left[\epsilon_i - \epsilon_{i+1} = \frac{F_{i+1}}{E_{i+1} w t_{i+1}} - \frac{F_i}{E_i w t_i} - \frac{1}{2R}(t_i + t_{i+1})\right] \tag{A.12}$$

At this stage, a closed-form solution for F and R is a rather formidable task which is best left to the computer. However, when the elastic moduli of all layers are set equal (which is often a reasonable assumption for III–V compounds), a general, closed-form solution is possible. In the three-layer case, after much tedious algebra, the following relationships are found:

$$1/R = t^3[6(t_1 t_2 \epsilon_{12} + t_2 t_3 \epsilon_{23} + t_1 t_3 \epsilon_{13})]^{-1} \tag{A.13}$$

$$F_1 = Ewt_1\left[\frac{t_2}{t\epsilon_{21}} + \frac{t_3}{t\epsilon_{31}} - \frac{1}{2R}(t_2 + t_3)\right] \tag{A.14}$$

$$F_2 = Ewt_2\left[\frac{t_1}{t\epsilon_{12}} + \frac{t_3}{t\epsilon_{31}} - \frac{1}{2R}(t_3 - t_1)\right] \tag{A.15}$$

$$F_3 = Ewt_3\left[\frac{t_1}{t\epsilon_{13}} + \frac{t_2}{t\epsilon_{23}} + \frac{1}{2R}(t_1 + t_2)\right] \tag{A.16}$$

where $t = t_1 + t_2 + t_3$; and $\epsilon_{ij} = \epsilon_i - \epsilon_j = -\epsilon_{ji}$, which is the fractional difference in the natural lattice parameter between the ith and jth layers. (Note also that $\epsilon_{ij} = \epsilon_{il} - \epsilon_{jl}$.)

The general form for N-layers can be induced from these equations as

$$R = t^3\left[6\sum_{i=1}^{N}\sum_{j>i}^{N} t_i t_j \epsilon_{ij}\right]^{-1} \tag{A.17}$$

$$F_j = wEt_j \sum_{i=1}^{N}\left[\frac{t_i}{t}\epsilon_{ij} + \frac{t_i \kappa_i}{2R}\right] \tag{A.18}$$

where

$$t = \sum_{i=1}^{N} t_i \tag{A.19}$$

$$\kappa_i = \begin{cases} +1, & i<j \\ 0, & i=j \\ -1, & i>j \end{cases} \tag{A.20}$$

The stress in any layer is given by a sum of the stresses due to the force F_j plus the bending stress. The first stress is

$$\sigma_{F_j} = F_j/wt_j \tag{A.21}$$

The bending stress may be written as

$$\sigma_{B_j} = \frac{E}{2R}\left(y - \frac{t_j}{2}\right) \tag{A.22}$$

recalling that it varies from compression at the bottom to tension at the top. The distance y is the position in the layer measured *from the bottom.* Thus

$$\sigma_j = \frac{F_j}{wt_j} + \frac{E}{2R}\left(y - \frac{t_j}{2}\right) \tag{A.23}$$

or

$$\sigma_j(y) = E_j\left[\sum_{i=1}^{N}\left(\frac{t_i\epsilon_{ij}}{t} + \frac{t_i\kappa_i}{2R}\right) + \frac{1}{2R}\left(y - \frac{t_j}{2}\right)\right] \tag{A.24}$$

$$R = t^3\left[6\sum_{i=1}^{N}\sum_{j>i}^{N} t_i t_j \epsilon_{ij}\right]^{-1} \tag{A.25}$$

where $\epsilon_{ii} = \epsilon_{jj} \equiv 0$.

APPENDIX B. TETRAGONAL DISTORTION IN CUBIC CRYSTALS STRAINED IN TWO DIMENSIONS

Assume that a lattice-mismatched epitaxial layer is deposited onto an (001) substrate such that the epitaxial layer does not have its natural lattice parameter (referred to here as a_{pow}). A uniform elastic strain ϵ_{el} will exist everywhere in the interfacial plane of the layer and the lattice parameter measured along directions in this plane will have a strained value a, given as

$$a = a_{\text{pow}}(1 + \epsilon_{\text{el}}) \tag{B.1}$$

However, planes parallel to the interfacial plane (i.e., along the ⟨001⟩ axis) will be strained (ϵ_p) in the opposite manner due to the Poisson effect and have the value c, defined as

$$c = a_{\text{pow}}(1 + \epsilon_p) \tag{B.2}$$

The relationship between these quantities is found from Hooke's laws as follows:

$$E\epsilon_{xx} = \sigma_{xx} - \nu(\sigma_{yy} + \sigma_{zz}) \tag{B.3}$$

$$E\epsilon_{yy} = \sigma_{yy} - \nu(\sigma_{xx} + \sigma_{zz}) \tag{B.4}$$

$$E\epsilon_{zz} = \sigma_{zz} - \nu(\sigma_{xx} + \sigma_{yy}) \tag{B.5}$$

where E is Young's modulus, ν is Poisson's ratio, and σ_{ij} is the stress in the ith plane in the jth direction. For a uniform two-dimensional applied strain (ϵ_{el}) in the xy plane:

$$\epsilon_{zz} = \epsilon_p \qquad \text{(Poisson strain)} \tag{B.6}$$

$$\epsilon_{el} = \epsilon_{xx} = \epsilon_{yy} \tag{B.7}$$

$$\sigma = \sigma_{xx} = \sigma_{yy} \tag{B.8}$$

$$\sigma_{zz} = 0 \tag{B.9}$$

Equation (B.3) and (B.4) yield

$$E\epsilon_{el} = \sigma(1 - \nu) \tag{B.10}$$

Equation (B.5) is then

$$E\epsilon_p = -\nu(2\sigma) \tag{B.11}$$

Thus,

$$\epsilon_p = \frac{-2\nu}{1 - \nu}\epsilon_{el} \tag{B.12}$$

Since $\nu \simeq \frac{1}{3}$ (for most III–V compounds[27]), Eq. (B.12) reduces to

$$\epsilon_p \simeq -\epsilon_{el} \tag{B.13}$$

Equation (B.1) then becomes

$$a \simeq a_{pow}(1 - \epsilon_p) \tag{B.14}$$

and since $\epsilon_p = c/a_{pow} - 1$ [from Eq. (B.2)], the interfacial lattice parameter a is given as

$$a = a_{pow}[2 - (c/a_{pow})] \tag{B.15}$$

REFERENCES

1. J. H. van der Merwe, *J. Appl. Phys.* **41**, 4725 (1970).
2. J. H. van der Merwe and C. A. B. Ball, in *Epitaxial Growth*, J. W. Matthews (ed.), Academic Press, New York (1975).
3. G. H. Olsen and H. C. Snyman, *Acta Met.* **21**, 769 (1973).
4. W. A. Jesser and D. Kuhlmann-Wilsdorf, *Phys. Status Solidi* **19**, 95 (1967).
5. J. W. Matthews (ed.), *Epitaxial Growth*, Academic Press, New York (1975).
6. A. Y. Cho, *J. Vac. Sci. Technol.* **8**, 531 (1971).
7. M. J. Stowell, in *Epitaxial Growth*, J. W. Matthews (ed.), Academic Press, New York (1975).
8. E. Peissker, P. Mossen, and H. Alexander, *Phil. Mag.* **7**, 1279 (1962).
9. Y. Tauri, Y. Komiya, and Y. Harada, *J. Electrochem. Soc.* **118**, 118 (1971).

10. G. H. Olsen, M. S. Abrahams, and T. J. Zamerowski, *J. Electrochem. Soc.* **121**, 1650 (1974).
11. J. J. Tietjen and J. A. Amick, *J. Electrochem. Soc.* **113**, 724 (1966).
12. V. S. Ban, *J. Crystal Growth* **17**, 19 (1972).
13. C. T. Foxon, J. A. Harvey, and B. A. Joyce, *J. Phys. Chem. Solids* **34**, 1693 (1973); R. F. C. Farrow, *J. Phys.* **D7**, 2436 (1974).
14. C. J. Nuese, M. Ettenberg, and G. H. Olsen, *Appl. Phys. Lett.* **25**, 612 (1974).
15. O. Madelung, *Physics of III–V Compounds*, Wiley, New York (1964); D. L. Kendall, *Appl. Phys. Lett.* **4**, 67 (1964).
16. H. Nelson, *RCA Rev.* **24**, 603 (1963).
17. J. M. Woodall, H. Rupprecht, and W. Reuter, *J. Electrochem. Soc.* **116**, 899 (1969).
18. H. Nelson, U.S. Patent No. 3,565,709, filed February 14, 1969; M. B. Panish, S. Sumski, and I. Hayashi, *Met. Trans.* **2**, 795 (1971).
19. H. F. Lockwood and M. Ettenberg, *J. Crystal Growth* **15**, 81 (1972).
20. M. B. Panish and M. Illegems, *Proceedings Third International Symposium on GaAs*, The Institute of Physics, London (1970), p. 67.
21. G. H. B. Thompson and P. A. Kirkby, *J. Crystal Growth* **27**, 70 (1974).
22. N. Nagai and Y. Noguchi, *Appl. Phys. Lett.* **26**, 108 (1975).
23. R. W. Hoffman, The Mechanical Properties of Nonmetallic Thin Films, U.S. AEC Contract No. AT11-1-623, Technical Report No. 82 (1975).
24. R. H. Saul, *J. Appl. Phys.* **40**, 3273 (1969).
25. S. Timoshenko, *J. Opt. Soc. Am.* **11**, 23 (1925).
26. G. H. Olsen and M. Ettenberg, *J. Appl. Phys.* **48**, 2543 (1977).
27. W. A. Brantley, *J. Appl. Phys.* **44**, 534 (1973).
28. M. S. Abrahams, L. R. Weisberg, and J. J. Tietjen, *J. Appl. Phys.* **40**, 3754 (1969).
29. H. Kressel, *J. Electronic Materials* **4**, 1081 (1975).
30. M. Ettenberg and R. J. Paff, *J. Appl. Phys.* **41**, 3926 (1970).
31. V. Swaminathan and S. M. Copley, *J. Am. Ceram. Soc.* **58**, 482 (1975).
32. A. G. Guy, *Introduction to Materials Science*, McGraw–Hill, New York (1972), p. 176.
33. I. Hayashi, M. B. Panish, P. W. Foy, and S. Sumski, *Appl. Phys. Lett.* **17**, 109 (1970).
34. W. A. Brantley, *J. Appl. Phys.* **44**, 535 (1973).
35. C. J. Nuese and G. H. Olsen, *Appl. Phys. Lett.* **26**, 528 (1975).
36. G. E. Fenner, *J. Appl. Phys.* **34**, 2955 (1963).
37. J. R. Biard, G. E. Pittman, and J. F. Leezer, *Proceedings 1966 International Symposium on GaAs* (unpublished), pp. 113–117.
38. R. L. Hartman and A. R. Hartman, *Appl. Phys. Lett.* **23**, 147 (1973)
39. M. Ettenberg and H. Kressel, *J. Appl. Phys.* **47**, 1538 (1976).
40. C. J. Nuese and J. J. Gannon, private communication.
41. J. Feinleib, S. Groves, W. Paul, and R. Zallen, *Phys. Rev.* **131**, 2070 (1963).
42. W. D. Johnston, Jr., and R. A. Logan, *Appl. Phys. Lett.* **28**, 140 (1976).
43. N. Nagai, *J. Appl. Phys.* **45**, 3789 (1974).
44. G. H. Olsen and R. T. Smith, *Phys. Status Solidi (a)* **31**, 739 (1975).
45. G. A. Rozgonyi, P. M. Petroff, and M. B. Panish, *J. Crystal Growth* **27**, 106 (1974).
46. P. Petroff, *Lattice Defects in Semiconductors*, Institute of Physics, London (1974), p. 73.
47. G. A. Rozgonyi and M. A. Afromowitz, *Appl. Phys. Lett.* **19**, 153 (1971).
48. H. R. Pettit and G. R. Booker, *Proceedings 25th Annual Meeting, Electron Microscope and Analysis Group*, Institute of Physics, London, Conference Series No. 10, Cambridge Univ. Press, London (1971), p. 290.
49. M. S. Abrahams and C. J. Buiocchi, *J. Appl. Phys.* **45**, 3315 (1974).
50. M. S. Abrahams, C. J. Buiocchi, and G. H. Olsen, *J. Appl. Phys.* **40**, 4259 (1975).
51. J. W. Matthews and A. E. Blakeslee, *J. Crystal Growth* **27**, 118 (1974).

52. D. Laister and G. M. Jenkins, *J. Materials Sci.* **8**, 1218 (1973).
53. M. S. Abrahams, Defect Characterization in III–V Compounds, Final Report on Contract No. N00019-72-0145, Naval Air Systems Command (1972).
54. D. W. Pashley, *Thin Films*, American Society of Metals, Metals Park, Ohio (1964).
55. F. C. Frank and W. T. Read, *Phys. Rev.* **79**, 722 (1950).
56. M. S. Abrahams, J. Blanc, and C. J. Buiocchi, *J. Appl. Phys.* **45**, 3277 (1974).
57. M. S. Abrahams, J. Blanc, and C. J. Buiocchi, *Appl. Phys. Lett.* **21**, 185 (1972).
58. D. Kuhlmann-Wilsdorf, *Physical Metallurgy*, R. W. Cahn (ed.), North-Holland Publ., Amsterdam (1970), p. 794.
59. J. Blanc, *Phil. Mag.* **32**, 1023 (1975).
60. G. H. Olsen, M. S. Abrahams, C. J. Buiocchi, and T. J. Zamerowski, *J. Appl. Phys.* **46**, 1643 (1975).
61. M. S. Abrahams, L. R. Weisberg, C. J. Buiocchi, and J. Blanc, *J. Materials Sci.* **4**, 223 (1969).
62. V. S. Ban, private communication.
63. M. Ettenberg, *J. Appl. Phys.* **45**, 910 (1974).
64. G. H. Olsen, *J. Crystal Growth* **31**, 223 (1975).
65. J. W. Matthews and A. E. Blakeslee, *J. Crystal Growth* **29**, 273 (1975).
66. F. C. Frank, *Proc. Phys. Soc. (London)* **A62**, 202 (1949).
67. W. A. Jesser, *Phys. Status Solidi (a)* **20**, 63 (1973).
68. J. Weertman and J. R. Weertman, *Elementary Dislocation Theory*, Macmillan, New York (1964), p. 189.
69. H. F. Lockwood, private communication.
70. S. Kishino, M. Oguima, T. Kajimura, and K. Kurata, *J. Crystal Growth* **24**, 266 (1974).
71. M. Ettenberg, C. J. Nuese, J. R. Appert, J. J. Gannon, and R. E. Enstrom, *J. Electronic Materials* **4**, 37 (1975).
72. M. S. Abrahams and C. J. Buiocchi, *J. Appl. Phys.* **36**, 2855 (1965).
73. P. Petroff and R. L. Hartman, *Appl. Phys. Lett.* **23**, 469 (1973).
74. R. E. Nahory, M. A. Pollack, and J. C. Dewinter, *Appl. Phys. Lett.* **25**, 146 (1974).
75. R. H. Saul, *J. Electrochem. Soc.* **118**, 793 (1971).
76. M. Ettenberg, S. H. McFarlane, and S. L. Gilbert, *Proceedings Fourth International Symposium on GaAs*, Institute of Physics, London (1972), p. 29.
77. P. Hirsch, A. Howie, R. B. Nicholson, D. W. Pashley, and M. J. Whelan, *Electron Microscopy of Thin Crystals*, Butterworth, London (1971), p. 119.
78. T. G. J. van Oirschot, G. A. Acket, and W. J. Bartels, *J. Appl. Phys.* **46**, 1893 (1975).
79. M. B. Allenson, P. G. R. King, M. C. Rowland, G. J. Steward, and C. H. A. Syms, *J. Phys.* **D5**, L89 (1972).
80. G. A. Antypas, L. W. James, and J. J. Uebbing, *J. Appl. Phys.* **41**, 2888 (1970).
81. R. E. Enstrom and D. G. Fisher, *J. Appl. Phys.* **46**, 1976 (1975).
82. D. G. Fisher and G. H. Olsen, unpublished.
83. A. A. Turnbull and G. B. Evans, *J. Phys.* **D1**, 155 (1968).
84. G. A. Allen, *J. Phys.* **D4**, 308 (1971).
85. C. J. Nuese, G. H. Olsen, and M. Ettenberg, *Appl. Phys. Lett.* **29**, 54 (1976).
86. C. J. Nuese, G. H. Olsen, M. Ettenberg, J. J. Gannon, and T. J. Zamerowski, *Appl. Phys. Lett.* **29**, 807 (1976).
87. J. J. Hsieh, *Appl. Phys. Lett.* **28**, 283 (1976).
88. R. E. Nahory, M. A. Pollack, E. D. Beeke, J. C. DeWinter, and R. W. Dixon, *Appl. Phys. Lett.* **28**, 19 (1976).
89. W. A. Nash, *Strength of Materials*, Schaums Outline Series, McGraw–Hill, New York (1972), p. 122.
90. F. K. Reinhart and R. A. Logan, *J. Appl. Phys.* **44**, 3171 (1973).

Chapter 2

Aspects of Silicon Epitaxy

J. NISHIZAWA

The vapor epitaxy of silicon is a process of considerable technological importance and as such has been the subject of much research (some of this was reviewed in the contribution to Volume 1 of this series by D. W. Shaw of Texas Instruments). However, several basic problems have still not been fully investigated and the present text outlines work carried out on some of these at Sendai in the groups at the Research Institute of Electrical Communication and at the Semiconductor Research Institute. Some of this has appeared in a variety of separate papers; the aim here has been to draw the varied threads together in the hope of presenting a more coherent interpretation of the way in which silicon epitaxy can be controlled. Most of our work has been concerned with processes based on chlorosilanes, and we have particularly investigated factors controlling growth rate, crystal perfection, and purity. The techniques developed in the course of this work could have important implications for device technology, a point illustrated by a description of their application in the development of a new solid state device, the static induction transistor, which is the solid state equivalent of the vacuum triode. The latter, while already of importance for power devices, also shows promise for integrated circuits, in which it already has shown operation powers of less than 0.1 nW and with an energy of 0.002 pJ per gate. The highest speed for operation obtained with the simplest processing, consisting of two diffusion times after one epitaxy and followed by making contact holes, is 3.7 nsec. There is obviously the promise of subnanosecond operation with improved structure, but with more complicated processing.

J. NISHIZAWA · Research Institute of Electrical Communication, Tohoku University, Sendai; and Semiconductor Research Institute, Sendai, Japan.

1. MECHANISM OF VAPOR-PHASE EPITAXIAL GROWTH OF SILICON CRYSTALS

Mechanisms of vapor-phase epitaxial growth of silicon crystals can be divided into the volume reaction (homogeneous) and the surface reaction (heterogeneous).

1.1. Volume Reaction in the Vapor Phase[1]

The simplest volume reaction occurs in the pyrolysis of monosilane:

$$SiH_4 \rightarrow Si + 2H_2$$

The reaction proceeds most rapidly at 900 to 1000°C. Fine particles of silicon generated by this reaction were observed by 100-kV electron microscopy, as shown in Fig. 1. The various polygons in this microphotograph correspond to projections of shapes of the particles; for example, the hexagons fall under the category of regular octahedral particles (B). Regular tetrahedra are observed in the form of single crystals free from defects (A). Truncated regular tetrahedra are also observed with a (111) twin boundary in the case of form C.

Fine crystalline particles are generated once the growth condition has been optimized, and their growth proceeds at a rate nearly proportional to the square root of the time taken to pass through the epitaxial reactor (see Fig. 2). This generation of particles occurs above a certain SiH_4 concentration which depends on the temperature (see Fig. 4). No nuclei of foreign particles larger than molecular size can be detected, while the smallest silicon particle observed has a diameter of 200 to 300 Å. It appears, therefore, that growth occurs predominantly without the need for nuclei of impurity molecules or atoms, i.e., by homogeneous nucleation.

After the nucleation of fine particles free of impurity atoms in the earlier stages of growth, the surface reaction may proceed on the surface of the fine crystalline particles by adsorption of reactants. Consequently, it is observed that the generated particles increase in size without changing their numerical density with time (see Fig. 3).

The nucleation density of these fine particles is almost proportional to the cube of the concentration of gaseous material, and changes exponentially with inverse temperature. These relations are shown in Figs. 3 and 4. As a result, the activation energy is estimated to be on the order of 24 cal/mol. It has not yet been confirmed what the intermediate products are in this volume reaction.

In the case of chlorosilanes, observations regarding volume reactions are summarized in Table 1.

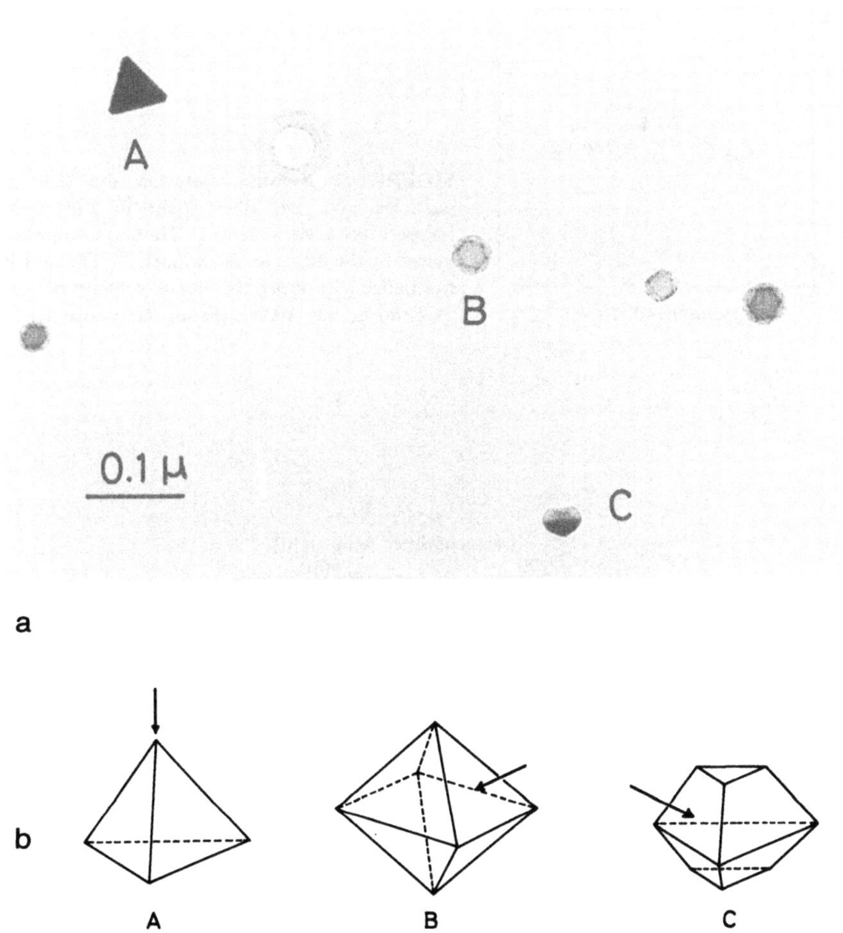

FIGURE 1. (a) Transmission electron microphotograph of Si particles formed at 1000°C with 0.2 vol% SiH_4 in H_2. (b) Equilibrium forms of fcc crystals which give the projected images shown in (a). Their surfaces are {111} facets. The arrow shows the direction of projection. (From Reference 1.)

1.2. Surface Reaction in the Vapor Phase

As the growth temperature is lowered, even in the case of pyrolysis of SiH_4, the volume reaction decreases remarkably and only the surface reaction occurs. Figure 5 shows the experimental data of Yasuda *et al.*[2] on polycrystalline deposition of silicon from SiH_4, where the growth rate on the *p*-type substrate is faster than that on the *n*-type substrate. (The growing layer doped with impurities acts as a substrate.) This suggests the occurrence of a catalytic reaction activated on the surface, and an

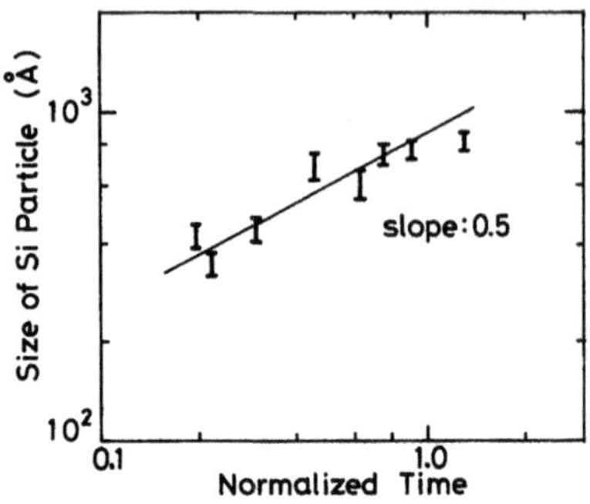

FIGURE 2. Relation between the size of particles and the time spent in the high-temperature zone at 1100°C. The size is approximated by the diameter of the particle. The time is normalized by using the linear velocity of gas, 20.7 cm/sec, at 1100°C. (From Reference 1.)

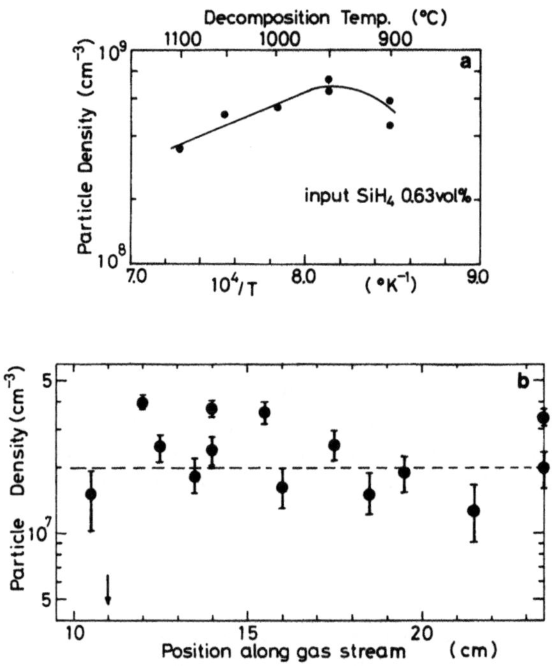

FIGURE 3. (a) Temperature dependence of silicon particle density. Input SiH_4 concentration, 0.63 vol%. (b) Dependence of particle density on the position along gas stream in the reactor, when the input SiH_4 concentration is 0.2 vol% and the temperature is 1100°C. The arrow indicates the position around which gas phase nucleation occurs. Particles are shown to be generated only near this point. (From Reference 1.)

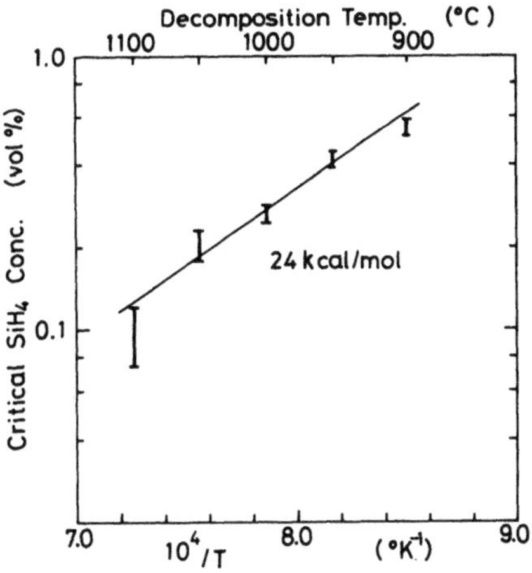

FIGURE 4. Temperature dependence of critical SiH_4 concentration above which Si particles are generated. (From Reference 1.)

importance for this catalytic reaction of the existence of holes which enables deposition to occur even at temperatures as low as about 250°C.

When silicon chlorides are used as source materials, the surface reaction predominates over the volume reaction and therefore allows one to observe only the surface reaction. (Details concerning the pyrolysis of SiH_3Cl or SiH_2Cl_2 are not available.) In a number of the investigations available in the literature,[3] however, it has been considered that the decompositions of $SiHCl_3$ and $SiCl_4$ can be described by the reactions:

$$SiHCl_3 + H_2 \rightarrow Si + 3HCl$$
$$SiCl_4 + 2H_2 \rightarrow Si + 4HCl$$

However, there is a great deal of uncertainty as to whether the growth rate of silicon is determined by an intermediate chemical reaction in the

Table 1. Occurrence of Volume Reactions in Silicon CVD

Type of silane	Concentration in H_2	Temperature (°C)	Formation of silicon powder
$SiCl_4$	less than 10%	<1250	not observed
$SiHCl_3$	less than 8%	<1200	not observed
SiH_2Cl_2	about 1–2%	<1050	observed
SiH_4	wide range		observed, as already published[1]

vapor phase, by a surface reaction, or by the migration rate of reactants on the surface. We have attempted to clarify the situation experimentally in a number of ways.

As part of our work we have studied the effect of ultraviolet irradiation. Ultraviolet irradiation of the reacting gas parallel to the silicon substrate but spaced from it by a few millimeters, does not affect the reaction rate at all. On the other hand, irradiation of the silicon crystal substrate promotes the growth rate in the irradiated region and also in a surrounding region to a distance of about 100 μm, indicating that the reaction is facilitated as a result of the activation of the surface by ultraviolet rays.[4] This appears to be a phenomenon related to the case of SiH_4 mentioned earlier, in which the growth rate is faster on a *p*-type substrate. Although the reaction temperature for the tetrachloride is about 1200°C, high-quality crystals, which contain hardly any dislocations, stacking faults, and other structural defects, can be obtained with uv irradiation at a slightly lower temperature, as shown in Fig. 6. The contribution of irradiation to growth rate is less at higher temperature, which may be caused by the increasing contribution of thermal energy to the migration of clusters adsorbed on the surface. The temperature rise, assuming that the irradiated ultraviolet rays are absorbed by the crystal, is found to be negligibly small.

This implies that the irradiation may contribute in any (or all) of three ways: to the generation of holes which promotes the surface

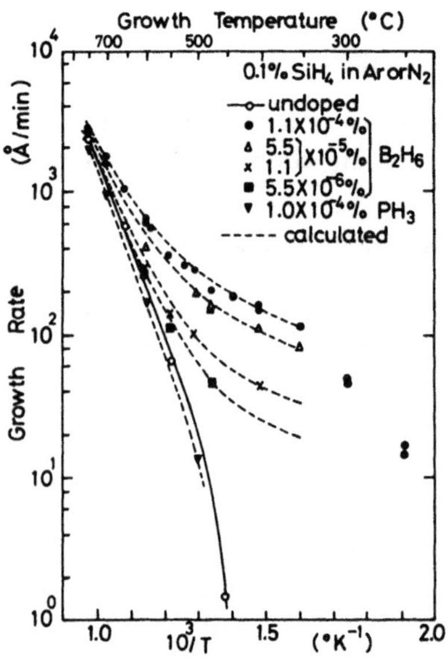

FIGURE 5. Temperature dependences of growth rates of undoped, and B- and P-doped polycrystalline layers. (From Reference 2.)

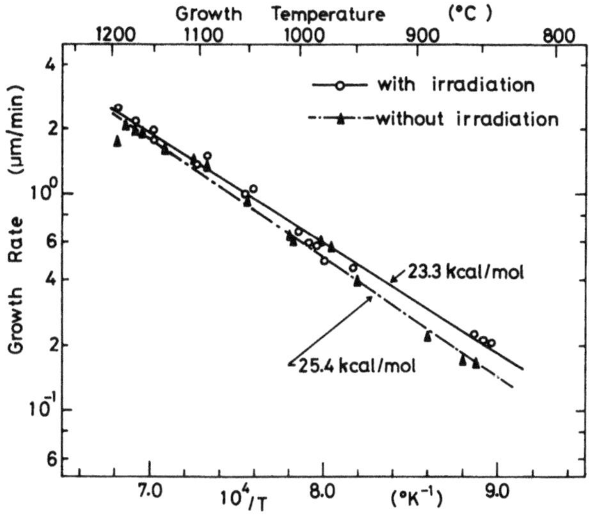

FIGURE 6. Temperature dependences of growth rates in epitaxial growth with and without ultraviolet irradiation in the $SiCl_4 + H_2$ system. Irradiating the surface of the substrate gives differences in growth rates and activation energies. Substrates are heated by rf induction. (From Reference 4.) (Reprinted by permission of the publisher, the *Japanese Journal of Applied Physics.*)

reaction, to the mobility of migrating adsorbed clusters, or to the occurrence of a photochemical reaction of intermediate products on the surface independently of the hole concentration.

Since the irradiation is effective in helping to grow crystals of good quality, it seems reasonable to postulate that the radiant energy furnishes an additional energy for migration of the clusters adsorbed on the surface to growth steps or kinks. These possible effects of irradiation may be further clarified in terms of the quantity of photon energy when more intense lasers are available in the future.

It is to be expected that undecomposed silane might not be adsorbed on a silicon surface in the absence of any reaction in the vapor phase. Now reactions in the vapor phase can be observed by various methods. For *in situ* observations, infrared spectroscopy is useful, but presents difficulties because of the high temperatures involved in the present case. Also, only the mean values of reactant concentrations for the entire path through which infrared rays pass can be obtained. As an alternative the wavelength and intensity of the light emitted on laser irradiation can be measured using the Raman effect, and the measured values used to determine the species, concentrations, and distributions of the intermediate products formed.[5] It is possible in this case to collect the Raman radiation from specific points along the path of the laser beam, thus enabling the spatial distribution of reactants to be determined. However, there is the possibility that the reactions would be accelerated under laser irradiation.

In contrast to the infrared spectroscopy mentioned earlier it is also possible by mass spectroscopy to analyze the composition of gases sampled from various places inside the reactor for determination of the reactant distributions, etc.[6] However, reactions can continue to occur during the extraction and measurement, and the reactants may undergo changes in concentration and chemical form with decrease in temperature. To allow for these effects, variation of concentration with time was measured by varying the extraction rate and thereby the corrected reactant distributions were obtained.

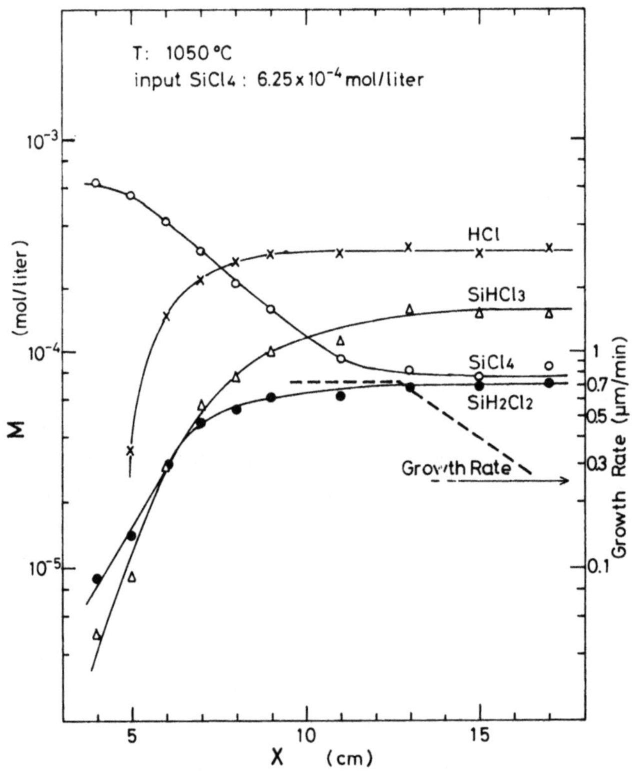

FIGURE 7. Distribution along the gas stream of concentrations of products from a mixture of $SiCl_4$ and H_2. Concentrations of product, *M*, are measured by ir spectroscopy of sampled gas at the position, *X*, from the beginning end of furnace. Growth rate of silicon is shown by the dashed line. (From Reference 6.)

Chlorosilanes, with the exception of monosilane, generally deposit silicon by a reaction with hydrogen and at the same time produce HCl. For example, $SiCl_4$ decomposes and produces several reactants in processes for the deposition of silicon, as shown in Fig. 7. This shows the concentration distributions of products from $SiCl_4$ along the gas stream in a horizontal reactor as measured by ir absorption spectroscopy of sampled gas. (These results can be extrapolated to give the contents in the reactor tube from the measurements as a function of flow speed in the sampling tube, which can be calibrated with the results of direct ir measurements through the reacting tube.) HCl, $SiHCl_3$, and SiH_2Cl_2 increase and then saturate in concentration as the gaseous mixture of $SiCl_4$ and hydrogen flows through the apparatus. This result indicates that the silicon deposition occurs in a complicated set of reactions between reactants and feed components in the Si–H–Cl system.

Conversely, the injection of HCl into the reacting system provides information on how spontaneously produced HCl changes the reactions. In Fig. 8, it is clear that a reverse reaction in addition to the regular reaction occurs on the injection of HCl into a quasi-equilibrium state, and $SiHCl_3$ and $SiCl_4$ concentrations increase as HCl concentration decreases. It is suggested from the results of Figs. 7 and 8 that $SiHCl_3$ (and SiH_2Cl_2) is produced in the process of deposition of silicon from $SiCl_4$, while backetching of silicon by HCl can also occur. There is experimental evidence, obtained by direct ir absorption spectroscopy through the high-temperature reacting gas in the tube, that $SiCl_2$ is also an intermediate reactant in addition to $SiHCl_3$ and SiH_2Cl_2 in the $SiCl_4$–H_2 system[6]; further details will be published.

1.3. Surface Migration

It is believed that the reactive intermediate products are produced only on the surface of the silicon substrate and form clusters which migrate on the surface. The presence of an adsorbed layer on the semiconductor surface has been ascertained in the case of gaseous diffusion.[45] This process is possibly promoted by the catalytic action of the silicon surface via the existence of holes or by uv irradiation, as mentioned in Section 1.2, because the reaction rate is decreased if the silicon substrate is not introduced and the intermediate activated substance is hardly produced. In the heteroepitaxial growth of silver or gold on NaCl, such a cluster migration can be observed by *in situ* electron microscopy. The clusters migrate in the form of islands possessing extraordinarily large migrating kinetic energies and successively unite with each other even though they may become separated from time to

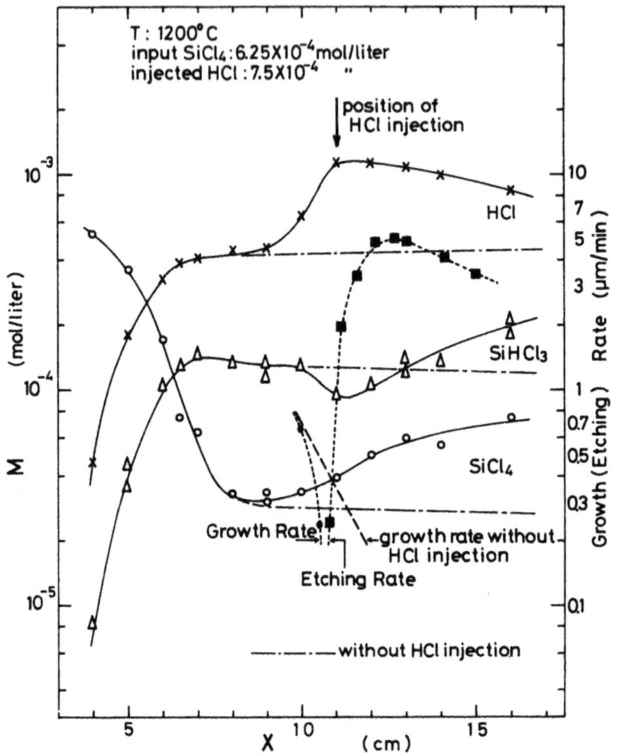

FIGURE 8. Distribution of concentrations of products from a gaseous mixture of $SiCl_4$, H_2, and injected HCl along gas stream. Growth rate and etching rate are shown by dotted lines. Dashed lines indicate behavior without HCl injection. (From Reference 6.)

time. When the clusters exceed a critical size, crystallization seems to be initiated inside the clusters.[7] It appears that a similar phenomenon may occur in the case of silicon. In our work a mixed gas of $SiCl_4$ and Pd-diffused H_2 was introduced to the heated substrate for several tens of seconds. The $SiCl_4$ was cut off; the substrate was then cooled and examined. It was found that when the quantity of the gaseous component exceeded a certain value, triangular pyramids were formed on the substrate surface (111) plane as shown in Fig. 9. The numerical density of these pyramids increased with increasing supply of the gaseous component and decreased with increasing temperature. The pyramid density, obtained as a function of temperature when the quantity of the gaseous component was constant, is shown in Fig. 10.

As in the case of the volume reaction (homogeneous nucleation, see Section 1.1) once the pyramidal hillocks have been formed, their density does not change so long as the temperature and supply rate of reactants are held constant; only their size increases.

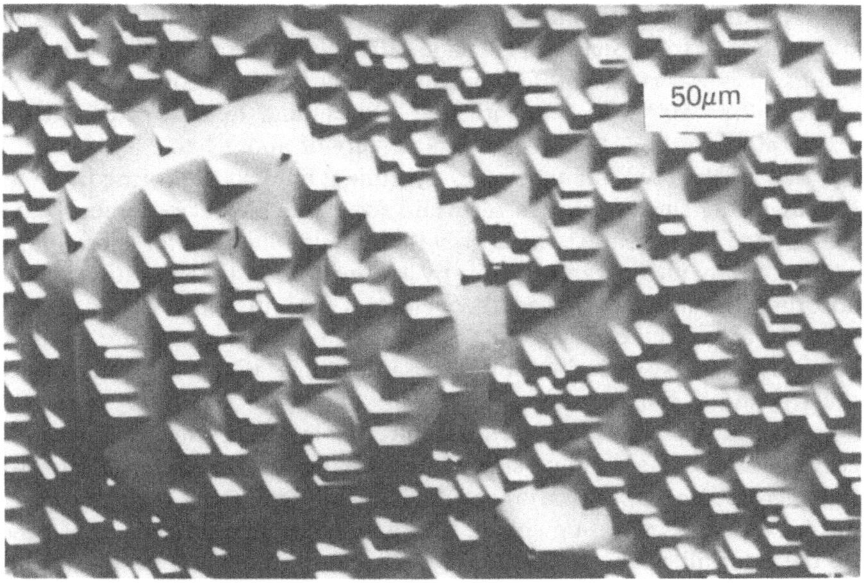

FIGURE 9. Microphotograph of pyramidal hills generated on a (111) facet by vapor growth at 1200°C for 30 sec followed by growth at 1000°C for 10 sec. The large circular table was prepared in order to measure lateral growth rate (see text). (From Reference 9.)

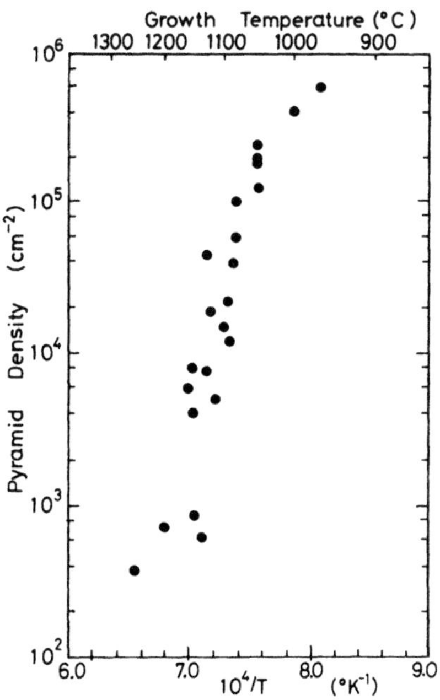

FIGURE 10. Temperature dependence of numerical density of pyramidal hills generated by vapor growth from 0.15% $SiCl_4$ and H_2. (From Reference 14.)

A similar experimental result has been published using molecular beam epitaxy from SiH_4 on (111) and (100) planes. From the experimental results, it is ascertained that when the quantity of evaporated molecules exceeds 2.7×10^{15} atoms/cm^2 at 830°C, pyramidal hillocks with flat tops are formed on the (111) surface at a density of about 6 to 7×10^5 cm^{-2}. In the case of the (100) plane, square pyramidal hillocks are formed.[8] When the migrating clusters exceed a certain size, each cluster becomes fixed to the surface and then acts as a growth nucleus. Such nuclei subsequently grow laterally into pyramidal hills by aggregation of the migrating clusters adsorbed around them. Adsorbed clusters present between two hillocks are not abundant enough for formation of new nuclei there, and the clusters which are adsorbed migrate and become attached to the edges of the pyramidal hills continuously. If the supply of $SiCl_4 + H_2$ mixture is abruptly increased or the temperature is lowered, the density which corresponds to the average size of adsorbed cluster is increased through the increase in concentration of the reacting materials or through the decrease of the migrating speed, and new pyramidal hillocks are formed on the interhill surfaces. Conversely, if the temperature is raised, the hills broaden to form flat tops and finally join each other because of the increase of the migrating speed of the adsorbed clusters. These results with the $SiCl_4$–H_2 system are similar to those found with molecular beam epitaxy.

The density of pyramidal hillocks and the lateral growth rate are shown as functions of the gas flow rate in Fig. 11. Lateral growth is observed parallel to the substrate surface exactly on (111) and this growth rate is measured as a spreading rate of a hillock or a propagation rate of a step on the exact (111) surface, i.e., the (111) facet, which can be as high as 100 μm/min under certain conditions. Eventually, the lateral growth rate becomes saturated with the increasing supply of reactant gas because the distance between the pyramidal hills, $2L$, is reduced and the absorbing reactant material to be cluster-migrating between hillocks is not increased even with the increase of supply. $2L$ is about 50 μm at 1200°C and 20 μm at 1150°C.

These hillocks are likely to be defect-free nuclei because they decrease in number reversibly with temperature and are almost homogeneously distributed on the surface. It appears that crystallization takes place when clusters adsorbed and migrating on the surface exceed the critical size. Although the critical size cannot be evaluated, the average effective density of adsorbed molecules is given by

$$-\frac{N_d}{\tau} + D\frac{\partial}{\partial x}\left(\frac{\partial N_d}{\partial x}\right) + v = \frac{\partial N_d}{\partial t} \qquad (1)$$

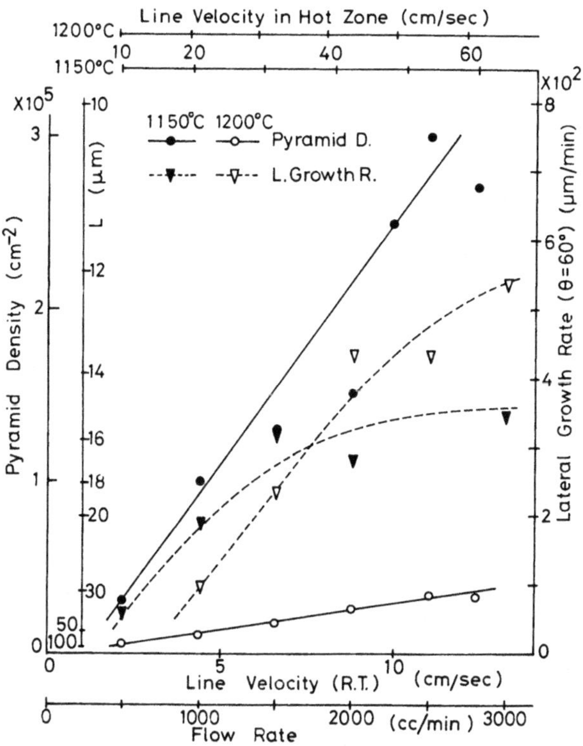

FIGURE 11. Linear gas velocity dependences of numerical density of pyramidal hills and lateral growth rate in the $\langle 1\bar{2}1 \rangle$ direction (for which the kink density is higher than that in any other direction). (From Reference 11b.)

where v is the particle density adsorbed from the vapor phase in unit time, τ is the average dwell time on the surface before reevaporation, and N_d is the surface density of particles which are adsorbed and migrate.[9] In Eq. (1) the stationary solution is approximated by putting $N_d = 0$ on the edge of hillocks as follows:

$$N_d \approx v\tau \left\{ 1 - \frac{\cosh[(x-L)/(D\tau)^{1/2}]}{\cosh[L/(D\tau)^{1/2}]} \right\}$$

$$N_d \approx \frac{v}{D} Lx - \frac{v}{2D} x^2 \qquad \text{when} \quad \tau = \infty \tag{2}$$

The growth rate is defined by

$$D\frac{dN_d}{dx}\bigg|_{x=0} = v(D\tau)^{1/2}\tanh\frac{L}{(D\tau)^{1/2}}, \qquad D\frac{dN_d}{dx}\bigg|_{x=0} \approx V(L-x)_{x=0} = vL \qquad (3)$$

The critical density N_c can be approximated by $N_d(x=L)$ (see Fig. 12). Consequently, DN_c is on the order of 9×10^{11} and 5×10^{11} atoms/sec at 1150 and 1200°C, respectively. Unfortunately these values cannot be directly compared with the experimental result of Joyce *et al.*[8] However, it is likely that the critical density in the $SiCl_4+H_2$ system might not be so very different from that in molecular beam epitaxy of SiH_4.

This description is the explanation as to how the clusters deposited on the surface are crystallized. Decreasing the supply of the gaseous material at elevated temperatures results in the activation of cluster migration, but hillock formation becomes difficult because of the low density of adsorbed clusters. As might be expected decreasing the supply of the gas has the disadvantage of lowering the growth rate.

When the density of adsorbed particles at $x=L$ is taken as N_c, the growth rate at $x=0$ and $2L$ as a function of the gas flow rate of 1.5 mol% $SiCl_4$ in H_2 is shown in Fig. 13. In our attempts to grow silicon without pyramid generation by the CVD (chemical vapor deposition) method on a Si substrate of 50-μm radius as an example, it was at least necessary to know the conditions where half the interpyramidal distance, L, is larger than 50 μm, the radius of the substrate (see Fig. 13a). At the same time, the growth rate is limited by the condition $L>50\,\mu$m, as shown in Fig. 13b. As a result, the maximum growth rate obtainable is limited to the range from 100 μm/min (experiment) to 200 μm/min (calculation) at 1200°C, and the linear velocity of the gas should be less than 20 cm/sec in the high-temperature zone of the reactor. At 1100°C, the growth rate is less than 40 μm/min and the linear velocity less than 10 cm/sec.

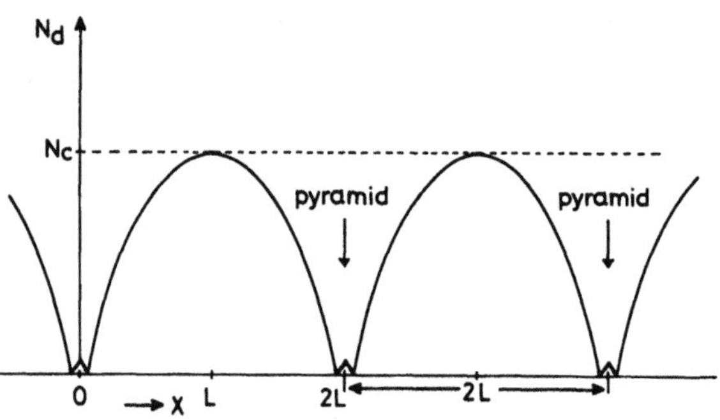

FIGURE 12. Schematic representation of density distribution of adatoms between pyramids. The adatom density at $x=L$ is approximated by the critical density N_c, above which pyramids are nucleated.

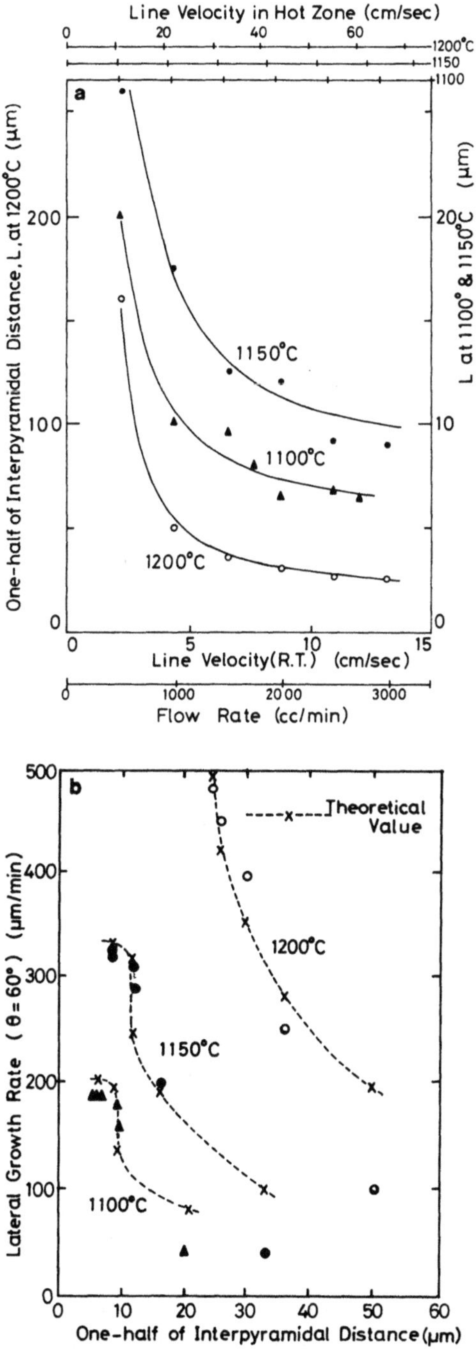

FIGURE 13. (a) The relation between linear gas velocity and interpyramidal (half) distance L, at 1200, 1150, and 1100°C. (Note different scale for 1200°C.) (b) L dependence of the lateral growth rate. Measured values at 1200°C (0), 1150°C (●), and 1100° (▲); ---×---- calculated values where $(D_\tau)^{1/2}$ is 20 μm at 1200°C, 12 μm at 1150°C, and 10 μm at 1100°C, respectively. (From Reference 11b.)

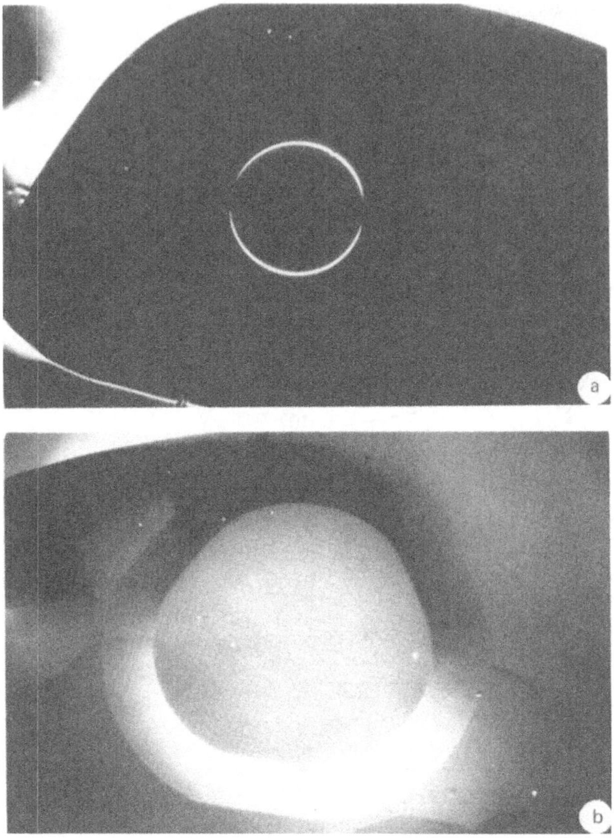

FIGURE 14. (a) Microphotograph (Nomarski method) of a round table with a diameter of 150 μm and a height of 1200 Å on the (111) facet, prepared by photolithographic techniques. (b) Microphotograph of the noncircular table produced from that shown in (a) after vapor growth at 1200°C for 60 sec. (From Reference 14.)

We prepared a substrate having a flat-topped hill in the form of a round table on the (111) plane, by the use of planar technology such as photolithography and oxidation.[10] An example is shown in Fig. 14a, taken by the Nomarski method. This substrate was made by first growing epitaxially on a dislocation-free and high-purity substrate cut very close to (111) surface orientation, by which an atomically flat facet was formed. This was followed by a first oxidation in steam and oxygen gas at 1200°C, photolithography using KPR (made by Eastman Kodak Company) as a photoresist, then selective etching of the oxide and a second oxidation also in steam and oxygen gas at 1200°C. During the latter the desired round table is formed by the oxidation of the bared silicon surface only, not of the still oxide-masked surface of what is to be the table. All oxide is then removed. This selective oxidation and oxide etching is required in order to obtain the required fine-dimensional tolerances.

Silicon grown by the CVD method on such a substrate shows a natural consequence that the clusters are best attached to kink sites,[10] as shown in Fig. 14b. However, since there is a considerable amount of step growth comparable to that of kink growth in all the experiments at the various growth temperatures, the energy difference between attachment at steps and at kinks seems to be very small. This, however, does not explain why the three corners which developed at the edge are rounded over a considerable length.

The shapes of hills which were grown on variously shaped substrates other than the circular hill just described are shown in Fig. 15.[11] It is evident that rounding of the corner does not occur in all cases, and

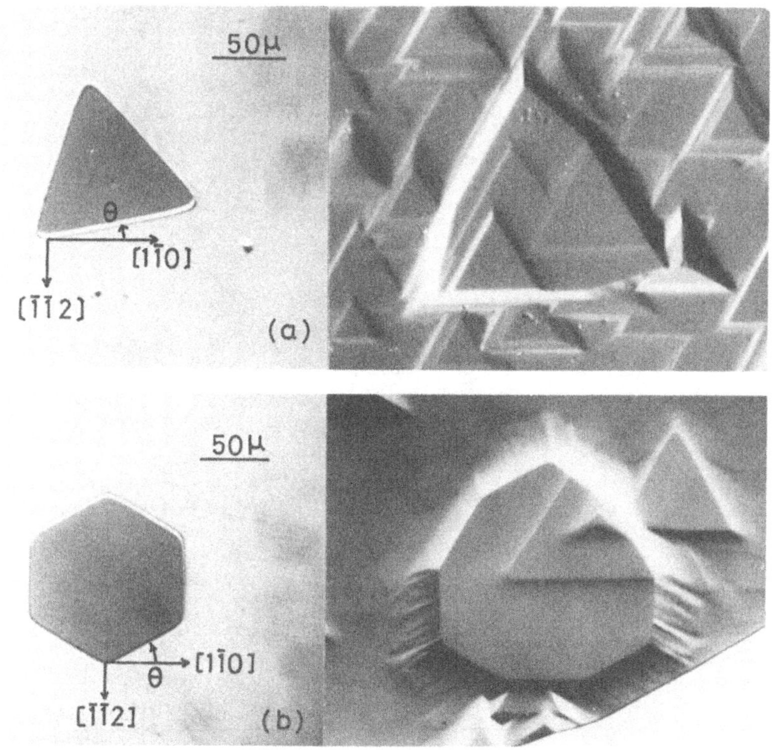

FIGURE 15. Microphotographs showing anisotropic lateral growth. (a) Deformation of a triangular table substrate into a hexagonal one during growth at 1100°C for 15 sec. The side of the original triangular table deviates by θ from the [1$\bar{1}$0] direction. (b) Deformation of a hexagonal table substrate into a nonagonal one during growth at 1140°C for 20 sec. (From Reference 11a.)

further inspection of Fig. 14 shows that the rounding is caused by the limited migration speed in the case in which the kink density is very high.

In short, it has become clear that the crystal growth rate, which has so far been considered as a whole, should be analyzed in terms of the supply rate of the gas, migration rate, kink density, step density, etc., as may be seen in Fig. 16. The lateral growth rate for all directions was measured using triangular tables formed on (111) planes, the sides of which were made to have different kink density as a function of θ (the deviation angle from the [110] direction on the (111) plane). The growth rate increased with the increase of gas flow rate (linear velocity) and with kink density (which is proportional to sin θ), although step growth was superposed on it. In the region of high kink density, however, the growth

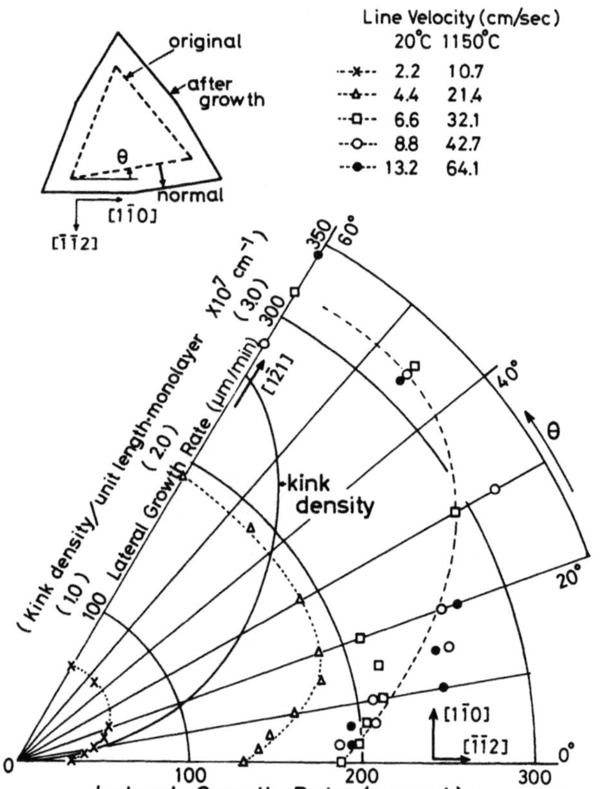

FIGURE 16. θ-dependences of kink density and lateral growth rate in the direction normal to a table edge measured as the spreading rate of a triangular table with various linear velocities of a 1.5% $SiCl_4 + H_2$ mixture. θ is the deviation angle of the side of a triangular table on the substrate from the [1$\bar{1}$0] direction. (From Reference 11b.)

rate is nearly constant independently of the density, which can be explained as due to limitation by the migration speed. The significant growth rate in the [112] direction $\theta = 0°$, for which kinks are scarce, is believed to be due to step growth caused by the attachment of migrating adatoms to step sites, which is consistent with the results just mentioned. This suggests that the lateral growth can be attributed to the (more favorable) attachment of adatoms to kink sites and by the significant (less favorable) attachment to step sites.*

The reason why substrates with surfaces inclined 3° off the (111) and (100) planes are used in practice is that the generation of pyramidal hillocks can be impeded by making a large number of kinks and steps on the substrate surface. By way of illustration a flat table prepared about 1° off the close-packed surface deforms to an inclined one during CVD as shown in Fig. 17, while many steps appear on the top surface.[12] When hillocks grow large, even a slight difference in lattice constant from that of the substrate can become accumulated on account of the misfit at the joins between hillocks which causes the generation of defects (in addition to those induced from the abnormally high kink and step density at the joins, which also produces an extremely inhomogeneous supply of migrating particles). In consequence, it is necessary for rapid epitaxial growth of silicon to use inclined substrates such as those 3° off the (111) plane. On such a substrate surface, stacking faults can sometimes be nucleated at foreign particles (oxide fragments or metal droplets) and at imperfections (e.g., a scratch). These show a (111) tetrahedral structure, which causes the appearance of a polygon on the grown surface, the shape of which is peculiar to the orientation, And then, if the growing layer meets with a stacking fault, much time is needed to cross over the stacking fault interface and thereby a step height difference is induced; that is, a depressed region is formed for each stacking fault, and the direction of this region corresponds to the projected layer growth direction as shown in Fig. 18.[13]

It needs comparatively large energy to cross over the interface of the stacking fault, simply because there are some irregularities in the arrangement of atoms at the interfaces, or at least a twin structure. Because of this, the layer speed to overwhelm the interface drops significantly compared with the usual rate of layer growth. This causes the depressions to be generated when a negative heat shock is given, because the energy needed to overwhelm the interface is larger and shows a larger dependence on temperature change. This energy depends on the interface structure of the stacking fault, e.g., twin lamellae. The growth rate over the interface seems to follow integral values, corresponding to the structure, e.g., to the number of twinned layers. Figure 19 illustrates these situations.[14]

*Similar ideas have been developed for solution growth by Shichiri *et al.*[10b]

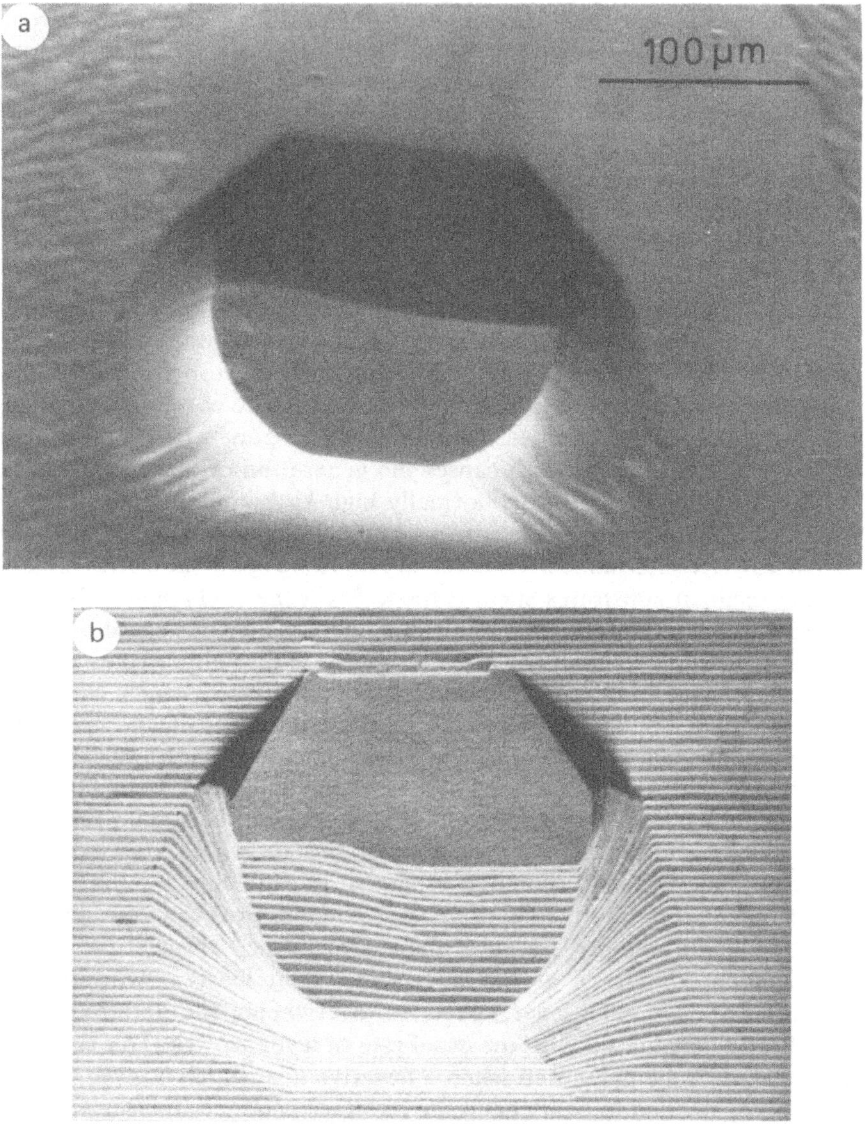

FIGURE 17. (a) Deformed table produced from a round one [prepared about 1° off (111) surface] by vapor growth at 1200°C for 60 sec. (b) Model of the deformed table shown in (a), representing the layered growth and the step structure.

FIGURE 18. Microphotograph of stacking faults and their shadows on an off-facet (111) surface taken by the negative phase contrast method. The depth of each shadow depends on the structure of the stacking fault joined. (From Reference 13.)

2. PERFECT CRYSTAL GROWTH FROM THE VAPOR PHASE

2.1. Perfect Crystal Growth on Exact (111) or (100) Surfaces

From the conclusions drawn from the investigations discussed in the preceding section, it is relatively easy to grow perfect crystals on the "3° off surface" since numerous steps are available to act as growth sites. Here, for convenience, crystals free from dislocations, stacking faults, and other induced defects except for point defects are defined as perfect crystal. However, there are many cases in which the (111) or (100) plane is required as the substrate surface. In this case, the vertical growth rate on the (111) plane is as slow as about 700 Å/min. As previously mentioned, if a negative heat shock is given, numerous new pyramidal hillocks are formed. If the growth temperature is restored, the pyramidal hillocks provide growth sites and commence to grow laterally, and eventually form a nearly atomically flat surface overall. The hillocks can then be regenerated by a second negative heat shock. In this way, it is possible to achieve reasonably rapid vertical growth on such a stable surface.

Furthermore, the quality of the crystal obtained is high enough to be free from dislocations, stacking faults, and other structural defects as determined from observations by optical microscopy and X-ray topography.

As an example, the shape of new hillocks is shown in Fig. 20. The angle a hillock surface makes with the crystal surface is very small, contrary to expectations, and was estimated to be 0.6 to 0.8° from the measurements of the size and the height of hillocks observed by the Nomarski method, which is similar to an oblique illumination, and by multiple reflecting beam interferometry which is sensitive to a surface relief as shallow as several angstroms due to very fine fringes.[15] (The

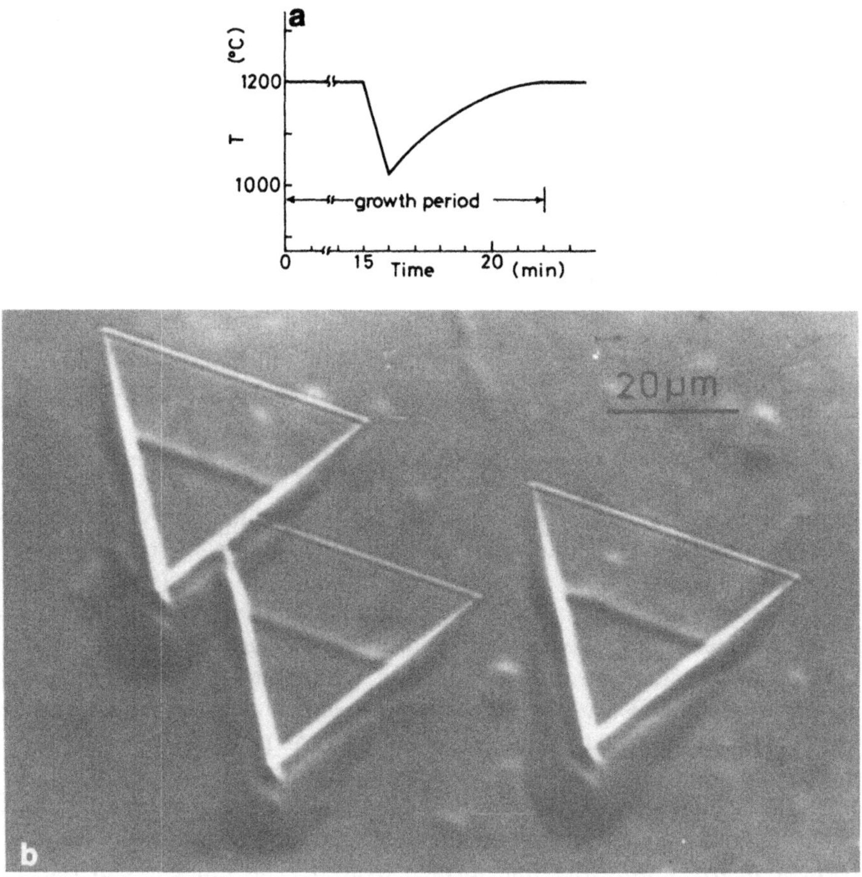

FIGURE 19. Microphotograph of stacking fault shadows (b), which are generated by a growth process with the temperature regime shown in (a).

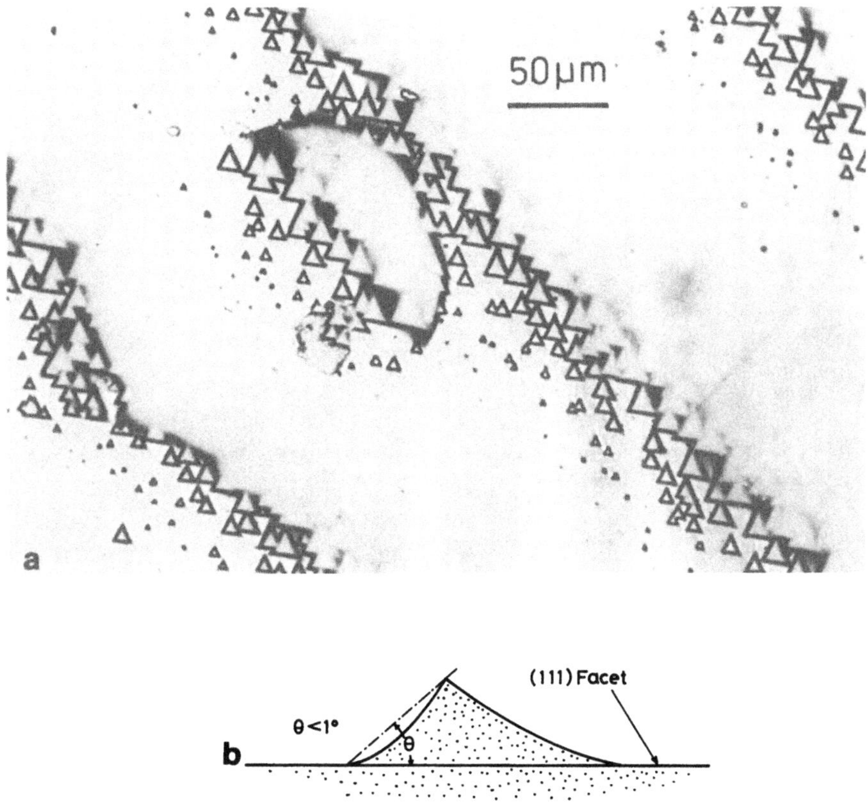

FIGURE 20. (a) Microphotograph of multiple-beam fringes of the pyramidal hills shown in Fig. 9. The spacing of fringes corresponds to a height difference of 2946 Å. (b) Schematic drawing of the cross section of a pyramidal hill deduced by means of interference contrast microscopy and multiple-beam interferometry. (From Reference 9.)

distance in the direction of illuminating light of two neighboring fringing lines on the crystal surface corresponds to half the wavelength of the illuminating light.) The ratio of the spread to the height gives a rough estimate of the angle of the relief. Figure 21 shows the smooth, perfect crystal surface of hillocks which have spread and joined each other.[15]

2.2. Growth of Highly Pure Crystals. Abnormal Impurity Distribution at the Substrate–Deposit Interface

In the growth of highly pure crystals, account has to be taken of backetching of neighboring substrates. Experiments in which a gaseous mixture of $SiCl_4$ and H_2 was reacted using substrates cut from two different kinds of doped silicon single crystals, gave the results summar-

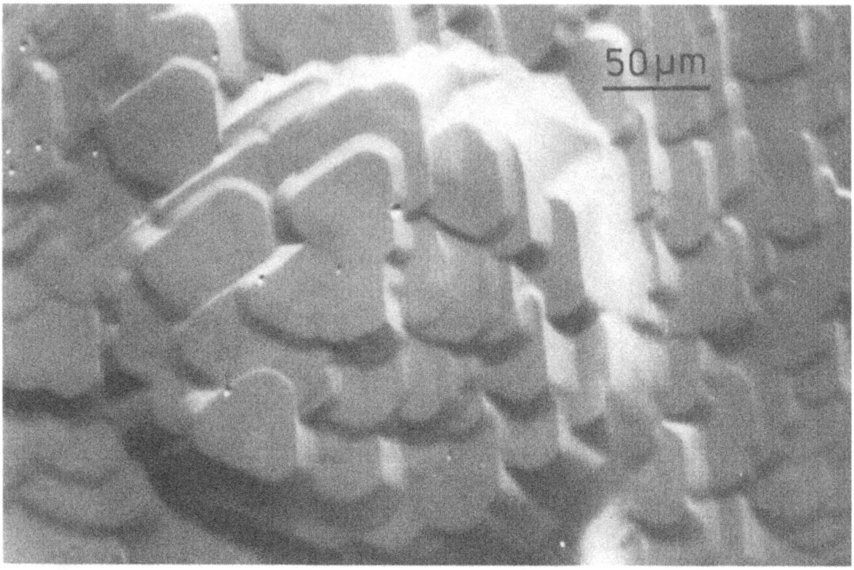

FIGURE 21. Microphotograph of pyramidal hills which are grown laterally and join each other by the consecutive vapor growth regimes: 1200°C for 30 sec, then 1000°C for 10 sec, and finally 1200°C for 60 sec. (From Reference 14.)

ized in Fig. 22. The purity of epitaxially grown crystals is mainly controlled by the contamination of the gaseous mixture supplied and by the substrate crystals. Especially in the case of heavily doped substrates, impurities are introduced into the grown layer by "autodoping" on account of backetching via the reverse reaction, as well as by the diffusion of doped impurity in the substrate toward the grown layer. The substrates used had dislocation densities less than $3000\,cm^{-2}$ normally and sometimes were nearly perfect when inspected following growth. The surfaces of substrates were cut off-orientation from (111) and were mechanically or chemically polished before the final treatment to be tested.

In order to investigate the effect of backetching, heavily doped and lightly doped *n*-type substrates were arranged alternately along the gas stream on the susceptor as illustrated in Fig. 22b, where the temperature distribution along the gas stream in the resistance heated reactor was as in Fig. 22a. As shown in Fig. 22c, the growth rate decreased with the change of the gas composition, as mentioned in Section 1.2, in a manner depending on the distance from the inlet to the furnace (to be more exact, on the distance to the top of the silicon substrate along the gas stream). The resulting carrier concentration profiles of layers grown on lightly doped substrates are shown in Fig. 22, where the thickness of the grown layers is normalized in terms of the growth time (because the growth

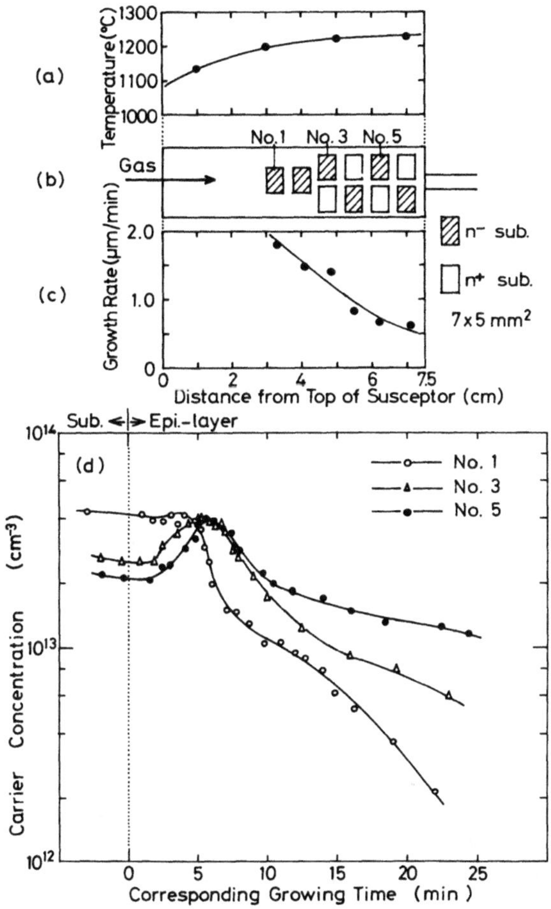

FIGURE 22. Experimental conditions to examine the effects of impurities in the substrate on the purity of epitaxial layers. (a) Temperature distribution in the reactor gas flow. (b) Arrangement of substrates. n^- sub. refers to phosphorus-doped material with a resistivity of 300 Ω-cm and n^+ sub. is antimony-doped material with a resistivity of 0.01Ω-cm. (c) Distribution of growth rate in the gas stream. (d) Distribution of impurity concentrations in epitaxial layers, obtained under the foregoing experimental conditions. The distance from the substrate surface (interface) is plotted versus the time of growth calculated from the growth rate shown in (c). (From Reference 16.) (Reprinted by permission of the publisher, The Electrochemical Society, Inc.)

thickness is a function of the composition of the reacting gas and of the growth time). It was found that high purity of the layer growing on the substrate is difficult to achieve at the backward (downstream) position. In addition, of course, impurities in the substrate tend to diffuse up toward the growing layer.[16]

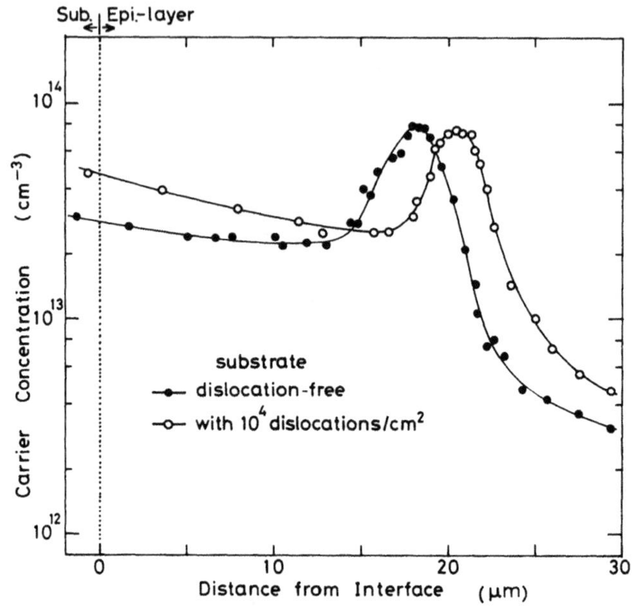

FIGURE 23. Impurity profiles of epitaxial layers grown on n^- substrate with dislocation densities of 10^4 cm^{-2} and zero. After the initial vapor growth for 20 min, the surface of the first epitaxial layer is exposed to the atmosphere for 24 hr, followed by the second growth for 20 min. Growth rate is 0.8 μm/min. (From Reference 16.) (Reprinted by permission of the publisher, The Electrochemical Society, Inc.)

When the first substrate has become covered with a highly pure grown layer and apparently no longer emits the impurity, i.e., after about 5 μm of growth, the second substrate commences to be covered with a more pure grown layer, and gradually the layers growing on substrates in the downstream positions increase in purity. These phenomena can be considered to be due to the effect of chemical backetching rather than to the evaporation of impurities. Therefore, for the growth of highly pure crystal, care should be taken not to have doped silicon nearby and to avoid the effects arising from backetching.

In a second series of experiments, after the growth of a first layer had been completed the substrate was taken out from the reactor and chemically etched. Then it was placed in the reactor again and a second layer grown. In this case a peak of carrier concentration appeared at the new interface as shown in Fig. 23. As can be seen, this was little influenced by the dislocation density at the surface.

Now, if growth of the second layer is performed without removing the substrate from the reactor, such a pronounced, abnormal peak in the carrier concentration distribution is not observed. However, when an intermediate heat treatment is performed in hydrogen still without removing the substrate from the reactor, an abnormal carrier distribution can be observed as seen in Fig. 24. This suggests that such abnormal distributions are caused by the presence of some kind of unstable layer on the crystal surface.[16]

In the case of the hydrogen reduction of $SiCl_4$, variation of growth temperature does not give any large difference in impurity distribution in the grown layer. In the thermal dissociation of SiH_4 over the temperature range of 1000 to 1200°C, however, reduced impurity concentrations seem to be found for nondoped layers grown at higher temperatures.[16] In any case, the epitaxial growth of a silicon layer which contains net active impurities less than $10^{13}\,cm^{-3}$ can in fact be achieved by the hydrogen reduction of $SiCl_4$, and this carrier concentration seems to be independent of the growth temperature and to be slightly dependent on the growth rate and on the input $SiCl_4$ concentration. This was only found possible,

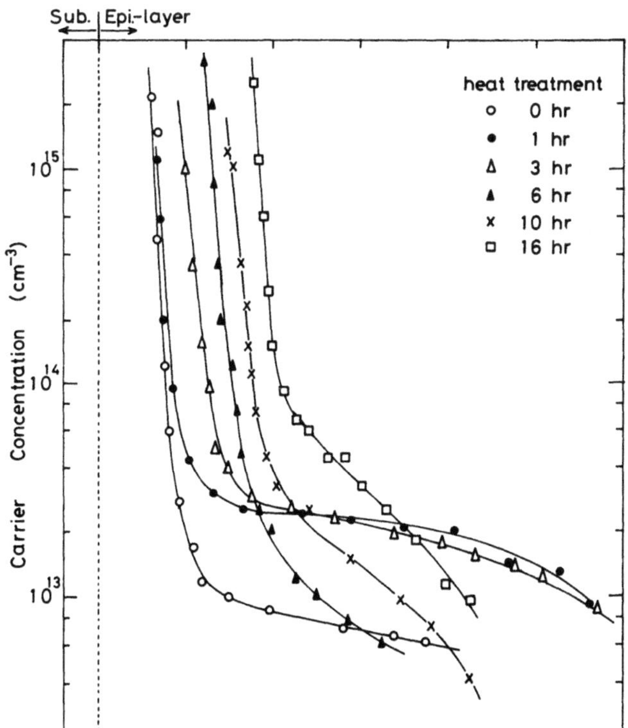

FIGURE 24. The variation of the carrier concentration distribution after heat treatment in H_2 at 1200°C for 0, 1, 3, 6, 10, and 16 hr. The epitaxial layer is 35 μm thick and grown for 30 min. (From Reference 16.) (Reprinted by permission of the publisher, The Electrochemical Society, Inc.)

however, when specially pure $SiCl_4$ and hydrogen, and an especially clean reactor were used. For this work, $SiCl_4$ was bought from the Toyo Silicon Company which was claimed to be so pure that the content of the residual impurities such as phosphorus, boron, and aluminum is less than 1 ppb. The hydrogen gas was commercially available and purified through a Pd diffuser, the residual impurities being believed to be water at less than 0.1 ppm. Before the experiment the reactor was baked at 1200°C in flowing hydrogen gas for about 10 hr. When the orientation of the substrate is exactly (111), a high growth rate can be obtained only with the negative temperature shock technique, which is also useful to obtain highly pure grown layers.

2.3. Effects in Silicon of Occluded Gas and Dissolved Material

Absorption of gaseous material into solids has been of considerable importance in various fields of engineering as in the case of hydrogen in iron, although it has attracted little attention in the field of semiconductors. It has long been established that in the case of oxygen absorbed in silicon, an infrared absorption at either 8.8. or 8.9 μm is observed depending on whether the oxygen is present in the form of single atoms or as SiO_2. Further, in the history of the development of the surface barrier transistor, it was found that solid germanium on heating dissolves in and diffuses into solid indium in contact with it, and as it does so, voids are formed at the interface between the indium and the germanium; these voids can be eliminated by saturating the indium with germanium beforehand.[17] (A possible explanation of this is that germanium in solution in indium is effectively in the metallic form and so has a considerably smaller radius than when tetrahedral.) In the case of silicon, account has been taken of such solid–solution in only a few papers[18] and the segregation of dissolved material from quenched silicon on heat treatment has usually been discussed in terms of foreign particles rather than voids. In any case the equilibria for solid–solute impurities can in practice be as important as those for impurities during growth at a solid–liquid interface and warrant further discussion here.

When the temperature is lowered, impurity atoms dissolved to saturation in silicon at high temperatures are often deposited when some nucleus exists within the diffusion length, or become frozen in when any nucleus does not exist within that distance. Such frozen-in materials frequently show retrograde solubility and can become nucleated or deposited on dislocations propagated during subsequent thermal treatment.

Normally the saturation solubility of carbon in silicon is 3×10^{18} cm^{-3}, beyond which value hexagonal SiC crystals are generated within the silicon. However, when antimony is present together with carbon, generation of SiC is not observable even if the carbon concentration reaches 7.6×10^{18} cm^{-3}, as shown in Fig. 25. This is an example of an induced stress altering the saturation solubility of carbon,[19] a matter discussed more fully in the following section. Suffice it to say here that antimony with its large radius tends to compensate for the lattice contraction induced by the small carbon atoms.

The amount of oxygen dissolved in silicon produced by the floating zone method is less than that for silicon obtained by other methods. This is partly because no SiO_2 crucible contamination occurs. However, what oxygen is present is depleted near the surface because of its evaporation in the form of volatile SiO. After cooling, oxide precipitation is also less near the surface.[20] Slow crystal growth rates tend to encourage precipitation of oxide and many clusters are formed, whereas at rapid growth rates oxygen is frozen and the generation of clusters or precipitates is scant.

It is usually assumed that oxygen, carbon, or metals which are frozen into silicon can in some cases become segregated and form clusters or swirls. Swirls can easily be revealed (e.g., by etching) when silicon vacancies and interstitials (or their complexes), which exist at high concentrations and do not give rise to observable impurity levels, become condensed on them. Such vacancy and interstitial depositions are also liable to occur when dislocations are propagated, or the degree of supersaturation of the point defects increases, or especially when heat treatment is performed.

In silicon crystals of 50-mm diameter produced by the floating zone method under atmospheres containing oxygen from material previously purified under vacuum, oxide deposits are distributed more on the surface and less in the center of the crystal, and the oxide concentration is higher in every section at a slower growth rate.[20] In this case, it is most likely that oxidation or dissolution of oxygen occurs at the surface and oxygen diffuses inward and after cooling forms precipitates at favorable locations because of the decrease of solubility.

It can, therefore, be stated that the concentration of gas dissolved in silicon crystals depends markedly not only on the method of production, e.g., the floating zone or the Czochralski techniques, but also on the atmosphere in which the crystal was grown. Similar behavior can also be found in epitaxially grown silicon, for instance, larger lattice constant, smaller breakdown field strength, and lower hardness,[21] which may perhaps be due to dissolved gaseous materials (hydrogen and chlorine

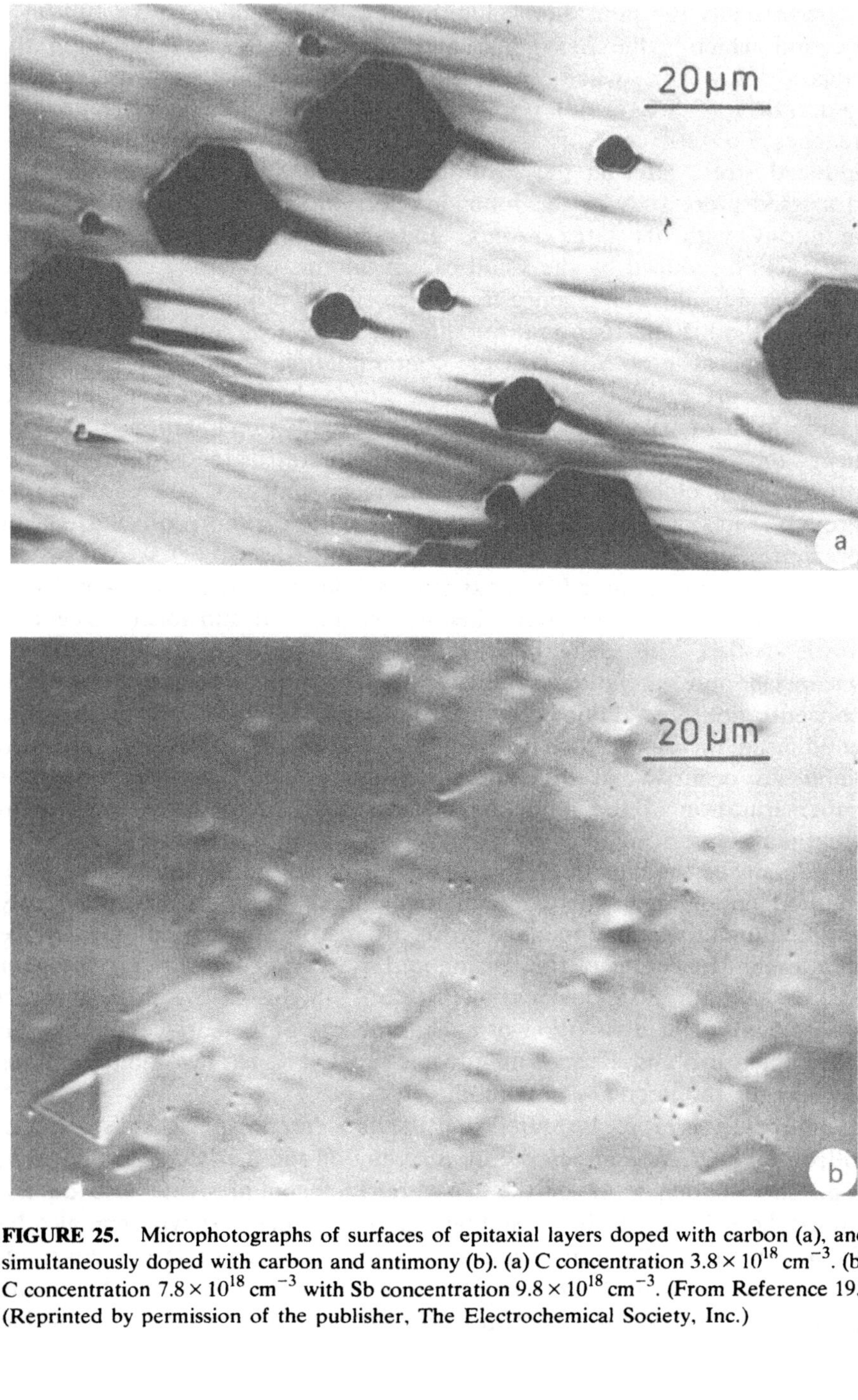

FIGURE 25. Microphotographs of surfaces of epitaxial layers doped with carbon (a), and simultaneously doped with carbon and antimony (b). (a) C concentration 3.8×10^{18} cm^{-3}. (b) C concentration 7.8×10^{18} cm^{-3} with Sb concentration 9.8×10^{18} cm^{-3}. (From Reference 19.) (Reprinted by permission of the publisher, The Electrochemical Society, Inc.)

have been shown to be present by Auger electron spectroscopy). A significant recovery toward normal values of these parameters can be achieved by heat treatment under vacuum, which encourages diffusion to the surface and subsequent loss of occluded gas, depending on its nature.

Much work remains to be done concerning the details of these phenomena. Generally, however, the absorbed gas and other solute materials which exist in large quantities in commercial silicon are normally supersaturated at room temperature but only tend to come out from the frozen-in state and segregate readily at a relatively higher temperature. Such segregation or precipitation is aided by the presence of lattice defects.

3. INFLUENCE OF EFFECTIVE LATTICE MISFIT

3.1. Change of Lattice Constant by Impurity Doping

The author, with others, published in 1953 that doping with an element in group IV of the periodic table enabled silicon crystal with high perfection to be made, and that silicon devices such as diodes with high breakdown voltage and low forward resistance could be realized.[22] These silicon crystals were prepared by a melt and freeze method. It is interesting to note that similar behavior was also observed with germanium tunnel diodes doped with indium. When the indium doping was increased, both the conductivity and peak-to-valley current ratio also increased up to a certain critical concentration of indium, beyond which they suddenly fell off. When the germanium was simultaneously doped with tin, however, these parameters showed no such tendency to fall off but continued to increase in identical manner to that for the case of lower indium doping.[23] Later, Schwuttke and co-workers found that the lattice constant of silicon is effectively changed by the diffusion of boron or phosphorus and that dislocations become generated.[24] Generally, if impurity atoms are substituted into silicon they push out or draw near the surrounding silicon atoms because the effective atomic radius of the impurity, R_i, is different from that of silicon, R_{Si}. The displacement is determined by the binding strength of the surrounding atoms. For example, atoms can undergo wide displacement near the surface because of the weak and asymmetric binding.[25] When the displacement, i.e., the differential strain, exceeds a certain value, a dislocation is introduced from the surface.[26] Macroscopically, the crystal is expanded or contracted by the introduction of impurity atoms as is shown in Fig. 26.

As a general guide the elements which possess a larger atomic radius

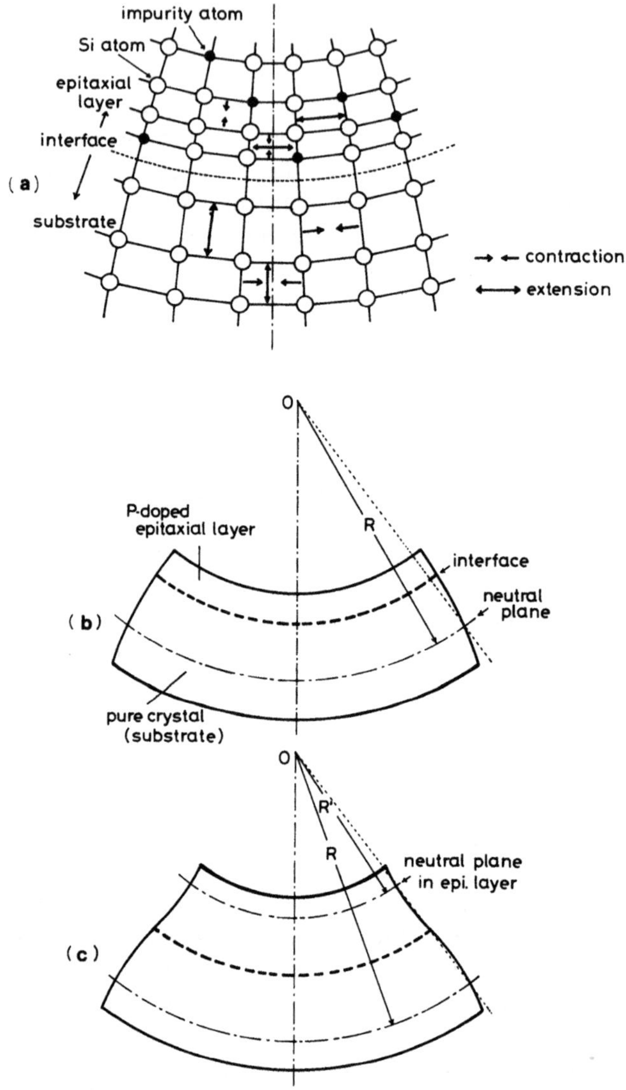

FIGURE 26. (a) Schematic drawing of the interface region of a bent crystal which comprises a substrate of pure silicon and an epitaxial layer doped with impurities smaller in atomic radius than silicon. (b) Outline drawing of the cross section of the bent crystal shown in (a) for the case in which the epitaxial layer is thinner than the substrate and the neutral plane exists only in the substrate. Below the neutral plane, the lattice of the substrate extends transversely and contracts longitudinally. (c) Outline drawing of the cross section of the bent crystal shown in (a) for the case of a thick epitaxial layer. Neutral planes exist in both the epitaxial layer and the substrate. The longitudinal lattice planes are deformed by lateral stresses. When the epitaxial layer is very much thicker than the substrate, the neutral plane exists only in the epitaxial layer.

than silicon are antimony and bismuth (donors), gallium, aluminum, indium, and thallium (acceptors), and, in group IV, tin and germanium (which are nondoping). Conversely, the elements which possess a smaller atomic radius than silicon are phosphorus (donor), boron (acceptor), and carbon in group IV. In addition, nitrogen as a donor[27] seems to possess a smaller atomic radius than carbon, while the atomic radius of arsenic is uncertain because of its complicated behavior which seems to depend, for instance, on its site in the lattice[28] or on cluster formation.[29] The tetrahedral radii of Huggins and Pauling[30] provide a good general indication of relative atomic size in this context, but the precise effective size in solution in silicon has to be determined experimentally, e.g., by high-precision lattice parameter measurements.

If a phosphorus-doped layer is grown on a pure silicon substrate, the resulting wafer bends as the thickness of the grown layer increases, on account of the difference of the effective lattice constants of substrate and epitaxial layer, as indicated in Fig. 26. The curvature, the reciprocal of the radius, also increases with increase in phosphorus concentration.[31] Conversely, if tin is introduced into the epitaxial layer, the wafer bends in the opposite sense because the atomic radius of tin is larger than that of silicon. These relations are plotted in Fig. 27. Bending of bicrystal is discussed in more detail in the following.

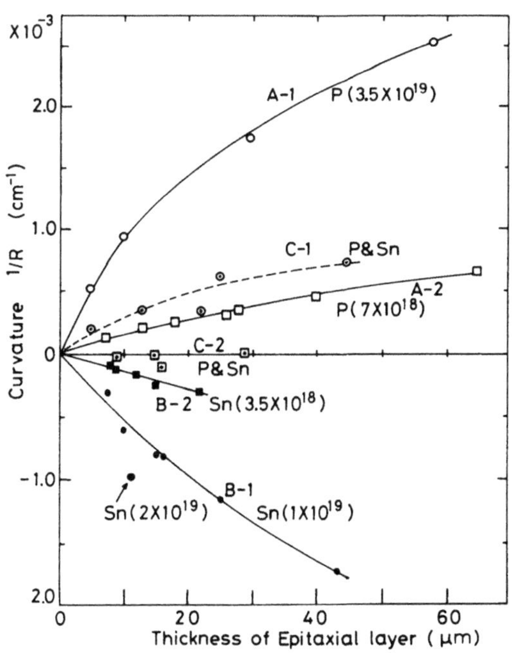

FIGURE 27. Layer thickness dependence of curvature of bicrystal comprising the substrate (~10 Ω-cm, boron-doped) 200 μm thick and the epitaxial layer doped with phosphorus (curves A-1, A-2) and tin (curves B-1, B-2), and simultaneously doped with phosphorus and tin (curves C-1, C-2). Curve A-1, phosphorus doping at 3.5×10^{19} atoms/cm^3. Curve B-1, tin doping at 1×10^{19} atoms/cm^3. Curve C-1, simultaneous doping of phosphorus and tin at 3.5×10^{19} and 1×10^{19} atoms/cm^3, respectively. Curve A-2, phosphorus doping at 7×10^{18} atoms/cm^3. Curve B-2, tin doping at 3.5×10^{18} atoms/cm^3. Curve C-2, simultaneous doping of phosphorus and tin at 7×10^{18} and 3.5×10^{18} atoms/cm^3, respectively. (From Reference 31.) (Reprinted by permission of the publisher, The Electrochemical Society, Inc.)

It should be noted that the silicon epitaxial layer containing phosphorus is stretched by the tensile action of the larger substrate lattice, which itself becomes contracted. It is significant that at the same time the epitaxial layer becomes contracted *in thickness* by an amount corresponding to this extension while the substrate material is correspondingly extended (Poisson's ratio effect). This behavior is illustrated in Fig. 26a and in terms of lattice planes in Fig. 28.

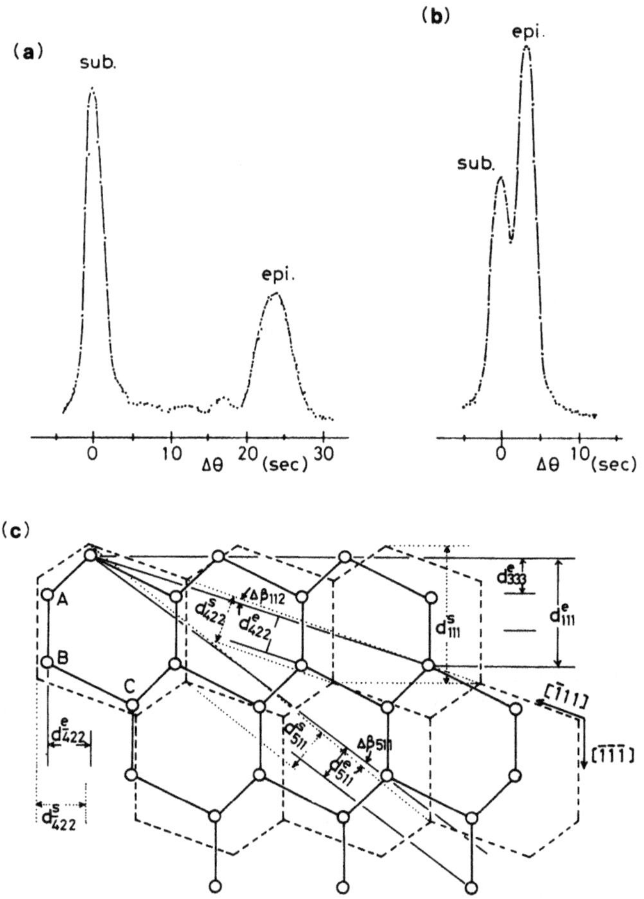

FIGURE 28. Strain situation near the center of bicrystal comprising P-doped epitaxial layer and pure substrate. X-ray rocking curves obtained by (a) $(511)^V$-$(333)^S$ and (b) $(511)^V$-$(511)^S$, respectively, using CuKα_1 radiation. (c) Schematic drawing of anisotropically deformed diamond lattice (solid lines) projected in the $[0\bar{1}1]$ direction, where open circles represent lattice points of the epitaxial layer and dotted lines indicate the diamond lattice of pure silicon free from strain. (From Reference 32.)

Since the strain is longitudinally symmetric in the plane of the interface because of the stress balance, few dislocations are induced. At the edge of the wafer, however, the behavior is somewhat different from that in the center. Approaching the substrate edge, the strain gradually becomes larger because of the asymmetry of stress and dislocations are easily induced.

We have examined the strain situation near the center of the epitaxial layer by the X-ray double-crystal method.[32] For the bicrystal comprising a phosphorus-doped epitaxial layer and a pure substrate with (111) surface, rocking angle curves were obtained as shown in Figs. 28a and 28b. The reflection planes employed were (333) (parallel to the surface) and (511) (having an intersecting angle of 38.9° with the surface), respectively. The two peaks in each curve are caused mainly by Δd, the difference in average spacing of each lattice plane in the epitaxial layer and in the substrate, although the lattice spacings gradually change near the bicrystal interface. Δd can be approximated as $-d\Delta\theta \cot \theta_B$, where $\Delta\theta$ is the angular separation between the two peaks and θ_B is the Bragg angle for both (333) and (511) reflections in silicon using CuKα_1 radiation. The angular separation, $\Delta\theta$, for the (511) reflection is smaller than that for the (333) reflection, which shows that the spacing of (511) planes in the epitaxial layer is more extended than that of (111) planes. As shown in the model of the strain situation in Fig. 28c, the spacing of the longitudinal lattice plane in the epitaxial layer is expanded by that for the substrate. Consequently, interatomic distance in the $[\bar{1}\bar{1}\bar{1}]$ direction, $\overline{AB}$, is reduced and that in the $[1\bar{1}\bar{1}]$ direction, $\overline{BC}$, is expanded, which leads to the result that the spacing between (111) lattice planes, d_{111}^e, increases more than that for the case of isotropic deformation. From the same relation, in the case of epitaxial growth on the (110) surface, the spacing between the lattice planes is shrunk in the [110] direction and extended in the [100] direction. As one moves away from the center of the substrate, the phenomena are not simple because the direction of stress is no longer symmetrical.

Furthermore, since the elastic constants of silicon are anisotropic, these relations are extremely complicated (see the related discussion by Olsen and Ettenberg in Chapter 1). However, the author believes the phenomena with silicon to be more straightforward than those with intermetallic compounds such as GaAs.

3.2. Compensation of Lattice Misfit

Various X-ray rocking curves of the epitaxially grown layers doped with phosphorus are shown in Fig. 29.[31] Each curve has two peaks, corresponding to diffraction from the substrate and from the epitaxial layer. It is obvious that the angular difference between the two peaks is larger with increasing phosphorus concentration and thickness of the grown layer, and also that the rocking curve gradually broadens, reflecting the deterioration in crystal quality. As shown in Fig. 30, X-ray projection topographs indicate also that the more heavily doped crystal is more inferior in quality and is bent more, and it can also be seen that the width of the crystal lying within the diffraction angle is narrower and the induced dislocation density higher.

Similar results in the case of tin doping are shown in Figs. 31 and 32. The shift of the peak in the rocking curve is opposite to that in the case of phosphorus doping. Figure 33 shows the experimental setup for a simultaneous doping by tin and phosphorus (and by other elements, as discussed later). Although the diffraction curves in the case of a single doping of phosphorus and tin have two peaks as shown in Figs. 34a and 34b, proper choice of the concentration ratio in such double doping gives only a single peak that coincides with the diffraction peak of the substrate. This coincidence is equivalent to the complete absence of lattice misfit. Moreover, it is clear from the sharpness of the peak that the crystal is of near-perfect quality (see Fig. 35).

Such an effect has been observed for many elements, in fair

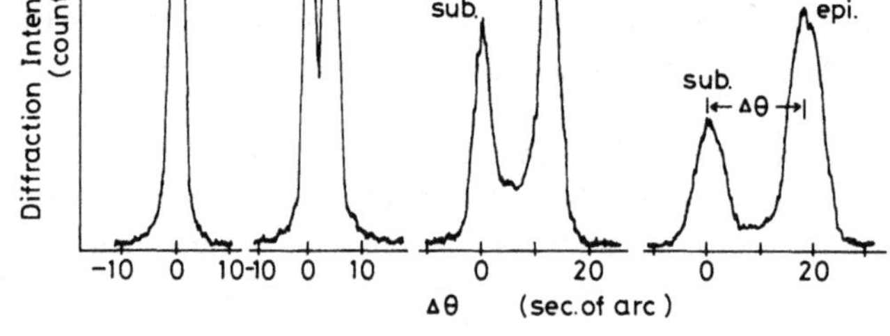

FIGURE 29. X-ray rocking curves of phosphorus-doped epitaxial layers on high-purity substrates. (a) Silicon substrate without epitaxial layer. (b,c,d) Bicrystals comprising substrates and epitaxial layers doped with phosphorus at concentrations of 7×10^{18}, 2.2×10^{19}, and 3.5×10^{19} atoms/cm^3, respectively. The thickness of each epitaxial layer is 13–15 μm, above which the diffraction intensity from the substrate decrease further. (From Reference 31.) (Reprinted by permission of the publisher, The Electrochemical Society, Inc.)

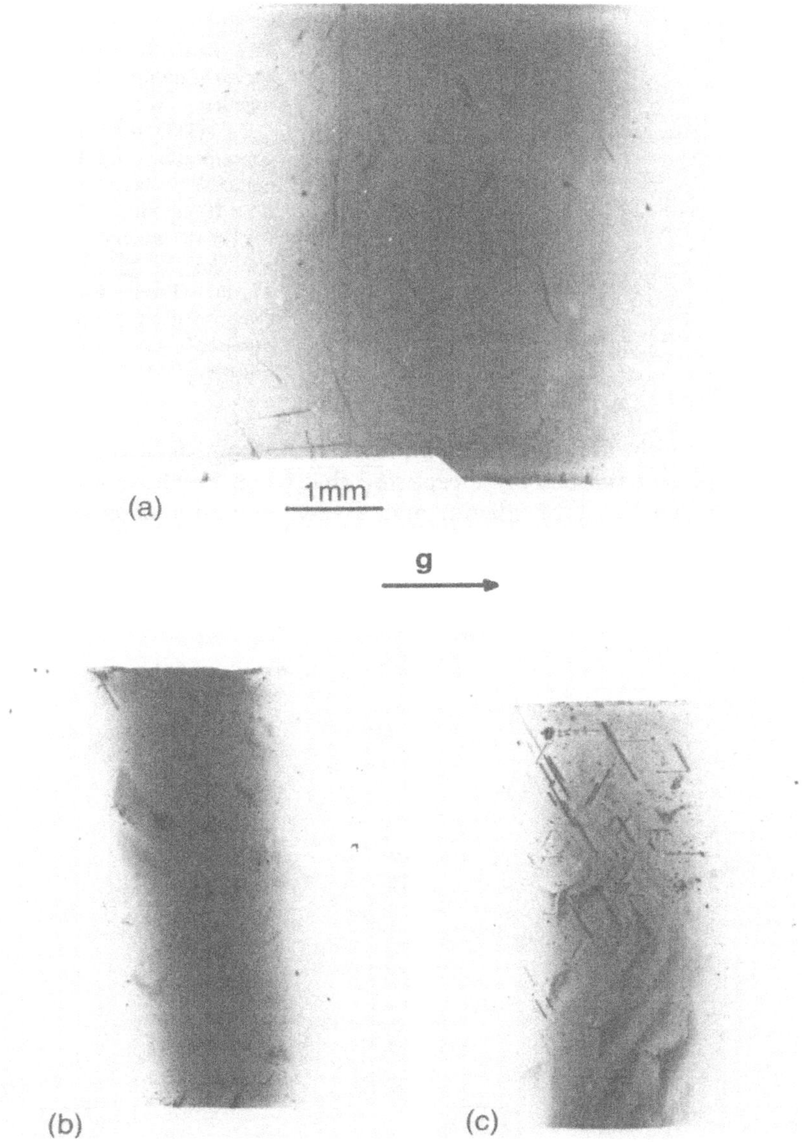

FIGURE 30. X-ray transmission topographs of phosphorus-doped epitaxial layers and substrates, where the thicknesses of the epitaxial layers are (a) 4 μm, (b) 10.5 μm, and (c) 14 μm, respectively, and the phosphorus concentration is 4×10^{19} atoms/cm^3. The diffraction plane used is (220), marked as **g**. The narrower the exposed region obtained, the more the crystal is bent. (From Reference 31.) (Reprinted by permission of the publisher, the Electrochemical Society, Inc.)

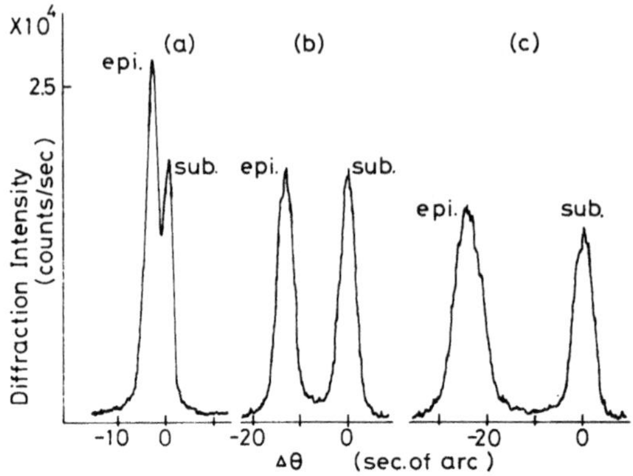

FIGURE 31. X-ray rocking curves of tin-doped epitaxial layers together with substrate by $(511)^V$–$(333)^S$ reflections. Impurity concentrations and thicknesses of epitaxial layers are (a) 3.5×10^{18} atoms/cm^3 and 15 μm, (b) 1×10^{19} atoms/cm^3 and 12 μm, and (c) 2×10^{19} atoms/cm^3 and 17 μm. (From Reference 31.) (Reprinted by permission of the publisher, the Electrochemical Society, Inc.)

agreement with the results estimated from their ordinary tetrahedral radii. Thus Watanabe *et al.*[33] reported that when P, whose atomic radius is smaller than that of silicon, and a small quantity of arsenic, which apparently is a little larger in atomic radius than silicon, are simultaneously added as dopants, strain compensation occurs at a P : As ratio of 3 : 1

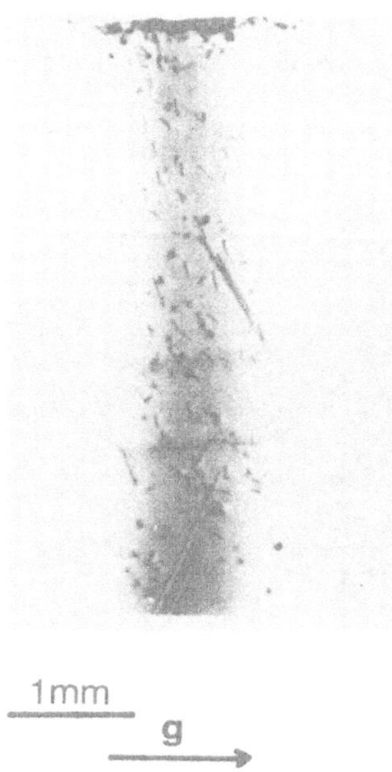

FIGURE 32. X-ray transmission topograph of tin-doped epitaxial layer and substrate. Concentration of tin is 1×10^{19} atoms/cm^3 and thickness of the epitaxial layer is 43 μm. (From Reference 31.) (Reprinted by permission of the publisher, The Electrochemical Society, Inc.)

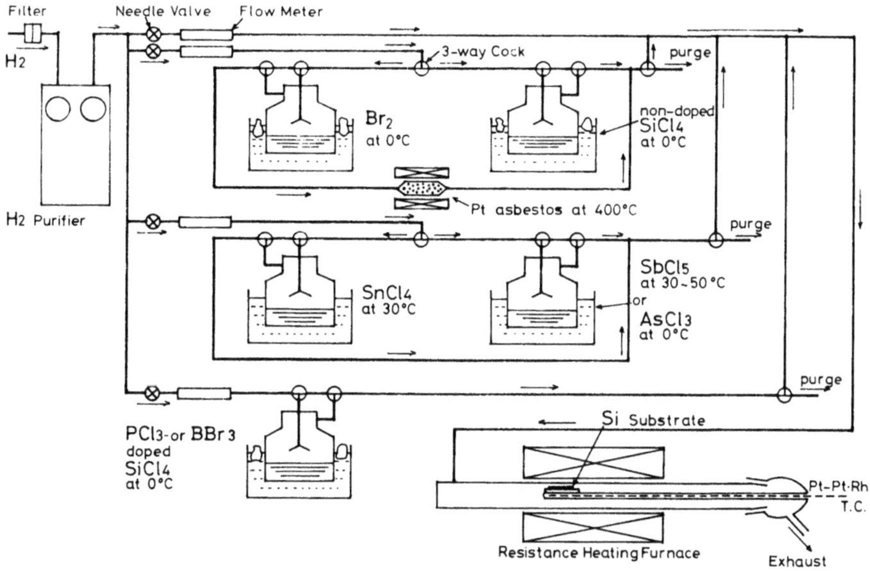

FIGURE 33. Schematic drawing of apparatus for vapor-epitaxial growth of silicon used for lattice-strain compensation. (From Reference 31.) (Reprinted by permission of the publisher, The Electrochemical Society, Inc.)

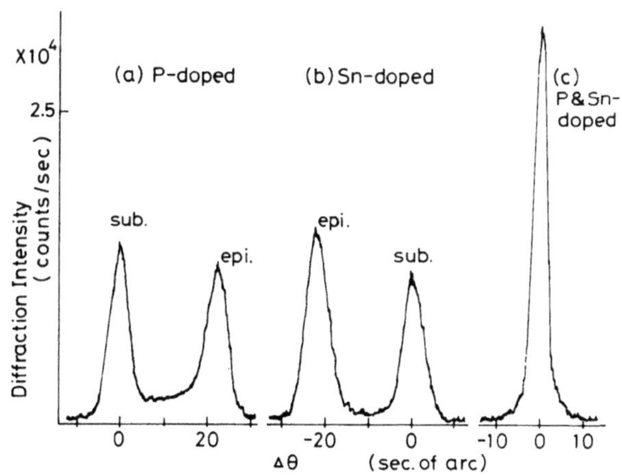

FIGURE 34. X-ray rocking curves of phosphorus-doped, tin-doped, and phosphorus and tin simultaneously doped epitaxial layers with substrate of pure silicon by $(511)^V - (333)^S$ reflection. (a) 4×10^{19} P/cm^3, epitaxial layer thickness 10 μm. (b) 2×10^{19} Sn/cm^3, epitaxial layer thickness 11.5 μm. (c) P and Sn simultaneous doping, 4×10^{19} P and 2×10^{19} Sn/cm^3, epitaxial layer thickness 16 μm. (From Reference 31.) (Reprinted by permission of the publisher, The Electrochemical Society, Inc.)

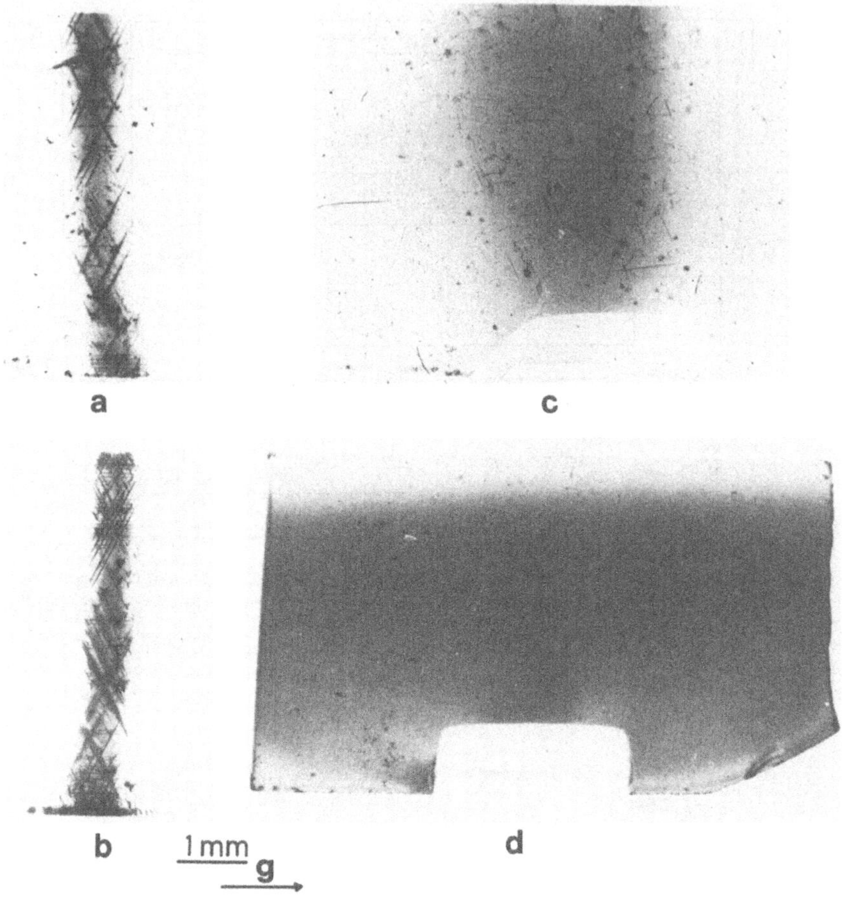

FIGURE 35. X-ray transmission topographs of phosphorus-doped and phosphorus and tin simultaneously doped epitaxial layers on pure substrates. (a,b) 3.5×10^{19} and 4.0×10^{19} P/cm^3, layer thickness 40 μm and 50 μm, respectively. (c) Simultaneous doping of phosphorus and tin. Concentrations 3.5×10^{19} P/cm^3, 1×10^{19} Sn/cm^3, layer thickness 43 μm. (d) Simultaneous doping of phosphorus and tin. Concentrations 4×10^{19} P/cm^3, 2×10^{19} Sn/cm^3, layer thickness 20 μm. (From Reference 31.) (Reprinted by permission of the publisher, The Electrochemical Society, Inc.)

to 4:1, which makes the crystal near-perfect. (As stated earlier although the behavior of arsenic is not yet understood, it has been suggested that it probably enters interstitial sites.[34])

Only carbon, tin, and germanium of the group IV elements can be used to bring about lattice compensation without producing electrically active impurity levels. (Titanium and lead do not seem to be suitable for this purpose from the author's experimental results, while potential applications of hafnium, zirconium, and vanadium for lattice compensation are still unknown.) The solubility of carbon in silicon has already been discussed in Section 2.3. There is a maximum solubility for tin of 5.5×10^{19} cm^{-3} at 1200°C, although this solubility limit may possibly be extended by doping with another smaller impurity such as boron, in much the same way as carbon doping affects the solubility of antimony, also mentioned in Section 2.3. Germanium by contrast shows complete solubility in silicon, which enables it to be used to compensate for large strain due to doping with impurities of radius smaller than silicon. According to our experiments, no reduction of bandgap is detectable up to a germanium concentration of 10^{20} cm^{-3}.(18) (It is interesting to note that according to some Soviet work, there is maximum solubility of boron in silicon of 2 or 3% at ~1300°C.(35) Presumably this could still be lattice-compensated for by an appropriate amount of germanium.)

Dislocations are only induced in silicon if a certain degree of localized stress is exceeded. This means that if it is only desired to suppress the formation of dislocations in, say, a semiconductor device with a boron-doped *p*-type region, then such perfect compensation as has so far been discussed may not be required. This opens up the possibility that adequate lattice compensation can be achieved by codiffusing appropriate elements, size and diffusion constant being the main factors in determining the codiffusant materials. In some of our experiments tin was first diffused into silicon, and then both tin and phosphorus were diffused simultaneously. This procedure was necessary because of the large difference of diffusion constants. Even so, the results showed that the generation of dislocations can effectively be impeded, although traces of the localized strain and stress still remain in the diffusion region because the concentrations of tin and phosphorus were not optimal(26) (see Fig. 36).

Effects of stress and strain on diffusion coefficients of impurities can be expected, although there are no reliable experimental data. However, it has been reported that abnormalities such as pileup layers generated near the surface in the single diffusion of phosphorus are suppressed in the double diffusion of phosphorus and tin,(26) and also in the simultaneous diffusion of phosphorus and arsenic (Fig. 37) where phosphorus atoms are distributed following a complementary error function.(33)

Backscattering measurements by the use of helium ions have suggested that arsenic atoms enter interstitial sites,(28) and that the fraction of such arsenic atoms codiffused with phosphorus is greater than that in the single diffusion of arsenic.(34)

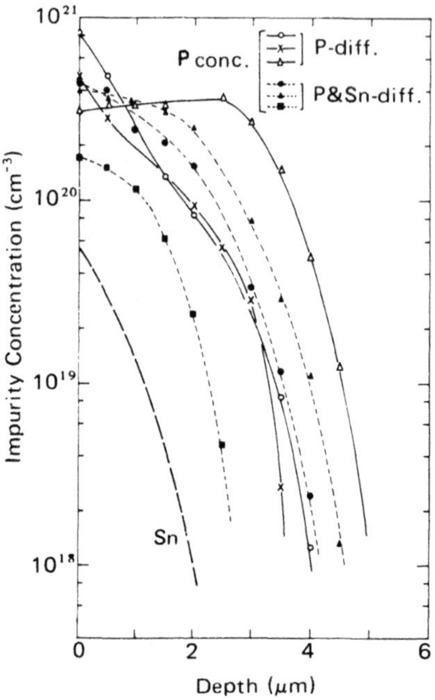

FIGURE 36. Diffusion profiles of phosphorus in the single diffusion of phosphorus and in the double diffusion of tin and phosphorus. Profile of tin is also shown as calculated values at diffusion temperature 1200°C for a period of 10 hr. (From Reference 26.) (Reprinted by permission of the publisher, the *Japanese Journal of Applied Physics*.)

4. DEVICE APPLICATIONS: THE STATIC INDUCTION TRANSISTOR AND INTEGRATED CIRCUITS

Vapor growth of high-perfection silicon device structures, free from lattice mismatch between high-resistivity pure regions and well-conducting highly doped regions, can of course be achieved by means of the techniques described earlier, and have important practical implications. This section outlines some of the work done in our group at the Semiconductor Research Institute in applying these techniques to the fabrication of a novel silicon power triode: the static induction transistor (SIT).

A transistor analogous to the vacuum triode was proposed by the author's group in 1950,[36] and also by Shockley in 1952.[37] However, such devices proved to be difficult to make; the first reasonably successful approaches were made by Zuleeg[38] and Teszner and Gicquel.[39] Their "analog transistor" devices did not in fact have characteristics like the vacuum triode.[40] The author's more recent development of the static induction transistor[41] has given an extremely practical device which, it turns out, is closely analogous in structure to the vacuum triode and has the same electrical characteristics (Fig. 38). It is in effect a power FET with an extremely small channel resistance (and hence, gate-source time constant), and a remarkably good frequency response.

Figure 39 gives a typical schematic representation of the SIT structure as used for low-frequency applications. A highly doped *n*-type substrate acts as drain. On this is epitaxially grown a 100-μm-thick, nearly intrinsic, high-purity layer ($\sim 10^{13}$ donors/cm^3). A grid gate of highly boron-doped, *p*-type material ($\sim 10^{20}$ B/cm^3) which also contains tin or germanium is then formed on this, followed by further epitaxial growth of the high-purity material ($\sim 10\ \mu$m thickness). Finally, a highly doped source region is deposited. For successful operation the gate and high-purity regions have to be lattice-parameter-matched by the technique described in the preceding section in order to avoid the formation of mismatch dislocations, which give rise to current leakage.

It should be noted that the near-intrinsic, high-purity layers (10^{13} cm^{-3} or less) which are mandatory in this type of device can be obtained by taking the precautions described in Section 2. Thus it is especially helpful for the growth to be carried out exactly on the (111) facet by the negative heat shock technique described in Section 2.1, and the precautions regarding purity discussed in Section 2.2 must be carefully observed.

Figure 40 shows a micrograph of the cross section of an actual device of this kind after stain etching. This is of *n*-channel type, with a (boron + tin)-doped gate grid. The grid lines are $\sim 25\ \mu$m in diameter on ~ 40-μm centers and the channels are $\sim 15\ \mu$m wide.

Crystal perfection within the device, as already mentioned, is particularly important, and there are two ways in which a lattice-matched grid

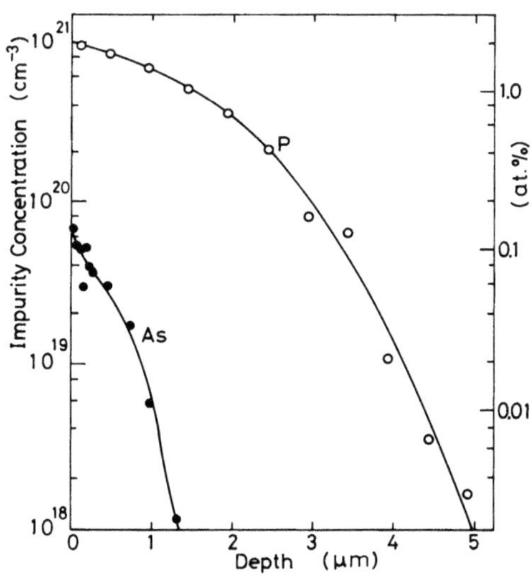

FIGURE 37. Concentration distributions of phosphorus and arsenic simultaneously diffused into silicon. Radioactive analysis. (From Reference 34.)

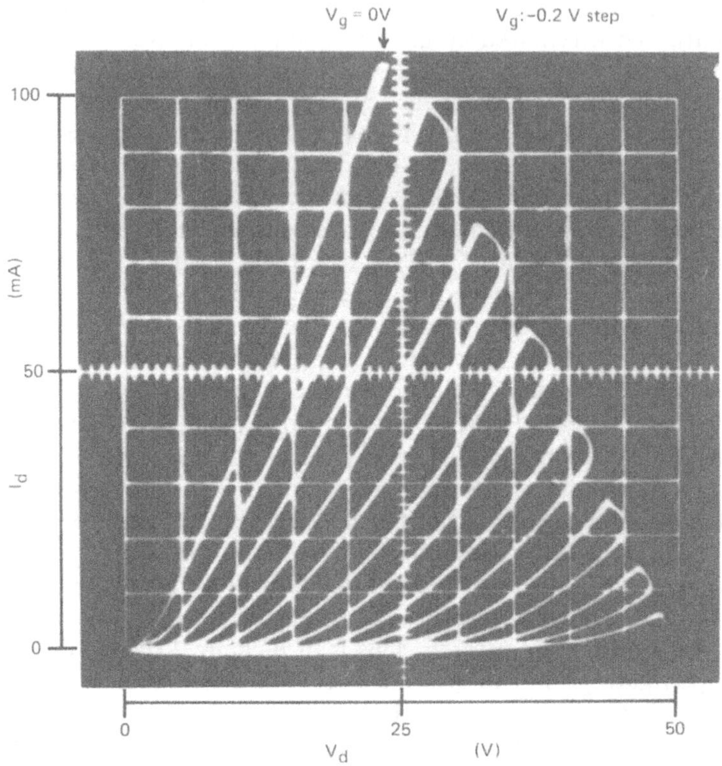

FIGURE 38. Typical $I_d - V_d$ characteristics of an SIT for various gate voltages. (From Reference 41.) (Reprinted by permission of the publisher, The Institute of Electrical and Electronics Engineers.)

can be formed; by vapor epitaxial growth and by selective diffusion. For the former a (boron + tin)- or (boron + germanium)-doped layer is grown on top of the first high-purity region and matched to it in lattice parameter as

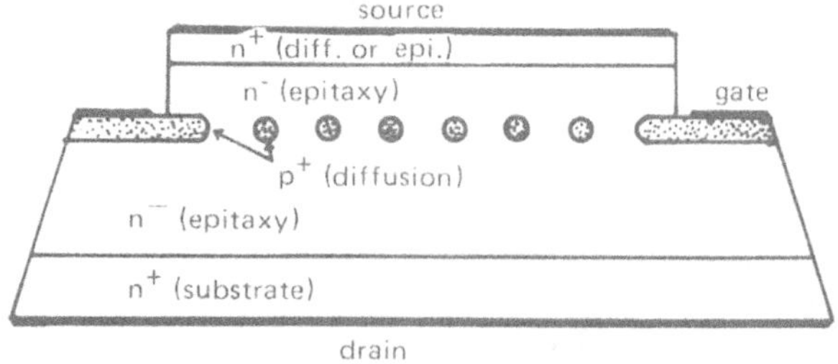

FIGURE 39. Schematic drawing of a cross section of an SIT. (From Reference 41.) (Reprinted by permission of the publisher, The Institute of Electrical and Electronics Engineers.)

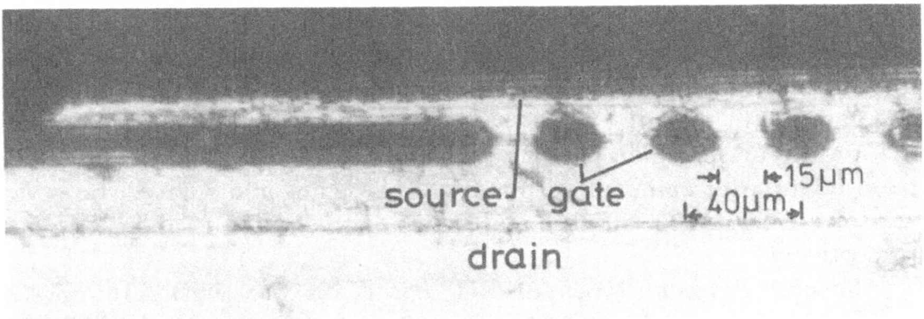

FIGURE 40. Microphotograph of a cross section of an *n*-channel SIT after cleaving and stain etching. (Original ×1000; figure reduced for reproduction purposes.)

described in Section 3.2. The grid pattern can then be formed in this layer by standard photolithography and etch techniques. This approach is useful to reduce the grid series resistance since the impurity concentration in the grid material is high and nearly constant across its cross section. Alternatively, selected area diffusion can be used (after oxidation and photolithography); first comes a prediffusion of tin at maximum concentration just insufficient to generate dislocations, followed by a double diffusion of boron + tin. (This two-step diffusion process is needed to allow for the more rapid diffusion of boron.) The structure at this stage is shown in Fig. 41. The oxide mask is etched off before the subsequent growth of the second high-purity region. Figure 42 shows an X-ray topograph of an experimental device taken at this stage in the process. It shows a dislocation-free structure which, however, is not completely lattice-parameter-compensated since a contrast pattern due to strain is produced in the gate region. (The complementary *p*-channel SIT is fabricated by corresponding processes in which the *n*-type grid is doped with arsenic only or with phosphorus + tin.)

Figure 43 shows the appearance of the top surface of an *n*-channel SIT structure after the final n^- layer has been grown, embedding the gate grid on high-purity material. The depressions, formed because of oxidation during the selective gate-grid diffusion, are somewhat rounded off by the epitaxial growth.

The two-step diffusion process offers certain processing advantages over the approach using epitaxial growth. Thus it permits higher resolution patterning and shorter channels, and it also gives higher breakdown voltages between gate grid and source or drain: Obviously, however, it can never give perfect lattice parameter compensation, which *is* possible via the all-epitaxial approach.

The SIT structure can be scaled up or down to cover a range of powers and frequency cutoffs, from 2 kW at a few megahertz to 10 W at

2 GHz.[42] Figure 44a presents a top view of a 2-GHz SIT and clearly indicates the surface morphology, seen in section and at higher magnification in Fig. 44b. This type of device has a modified geometry compared with that of Fig. 39 in order to minimize stray reactance. As before, the drain region is the n^+ substrate, while the channels are made from high-purity epitaxial n^- material. The gates and sources, however, are selectively double-diffused and lattice-compensated p^+ and n^+ regions, respectively.

Several different types of SIT based on our work are now in industrial production by Yamaha. Complementary n- or p-channel devices

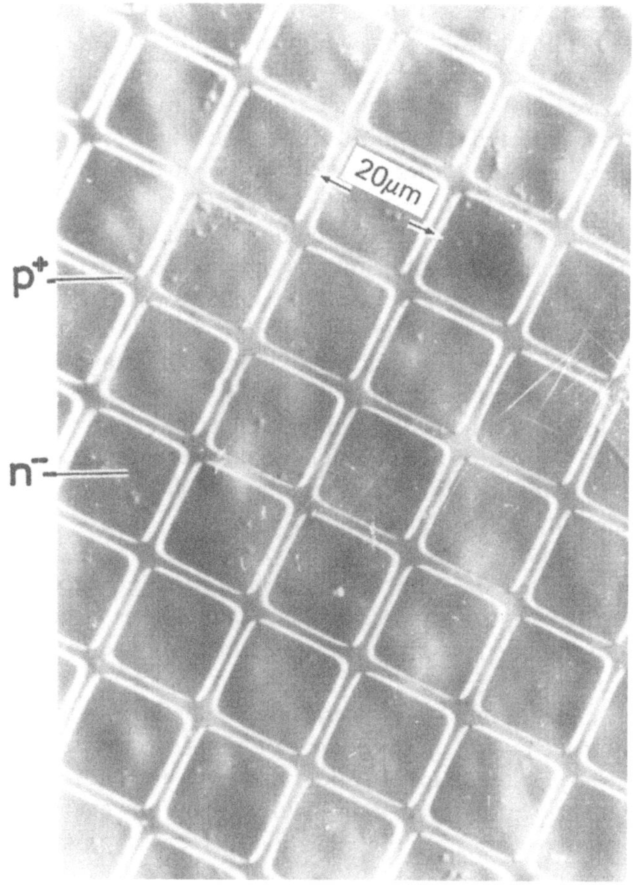

FIGURE 41. Microphotograph of lattice-patterned SiO_2 layer for the selective codiffusion of tin and boron for gate-grid fabrication.

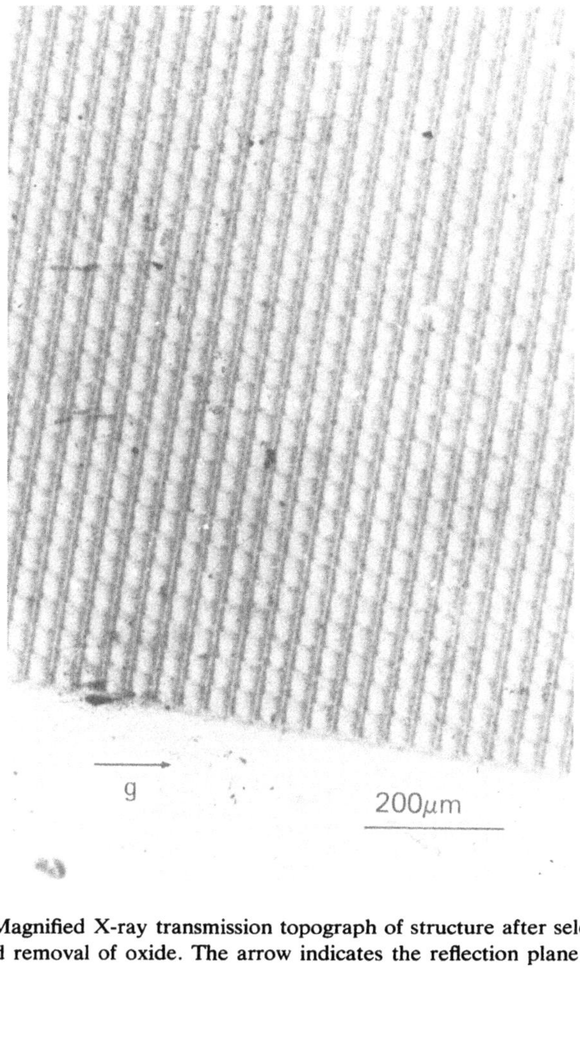

FIGURE 42. Magnified X-ray transmission topograph of structure after selective diffusion of gate grid and removal of oxide. The arrow indicates the reflection plane ($2\bar{2}0$).

are available and are useful for high-power operation, e.g., in audio-frequency amplifiers.

The SIT principle can also be applied with advantage to integrated logic circuits. Figure 45 shows schematically how such SIT logic (SITL) can be constructed. The output drain can be either a n^+ region or a Schottky contact. This kind of structure can be fabricated with only a three-mask planar technology. Such SITL offers high operating speed, very low power requirements (as small as 10% of that for normal bipolar I^2L, for example operation at 0.002 pJ/gate, or 30 pW dissipation has been demonstrated), as well as very low noise.

FIGURE 43. Microphotograph of surface of SIT structure after epitaxial growth of final n^- layer. (Original ×440; figure reduced for reproduction purposes.)

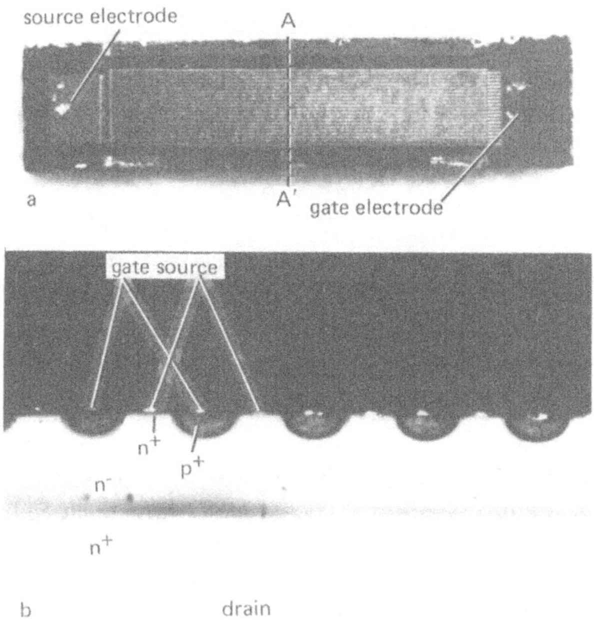

FIGURE 44. (a) Top view of structure of 2-GHz microwave SIT. (b) Microphotograph of cross section of microwave SIT after cleaving along line A–A and stain etching. (From Reference 42.) (Reprinted by permission of the publisher, The Institute of Electrical and Electronics Engineers.)

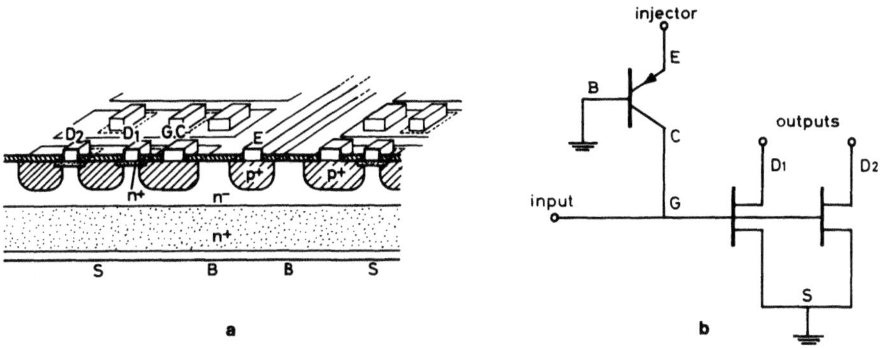

FIGURE 45. (a) Structure of I^2L type of integrated circuit based on SITL. (b) Circuit schematic for (a).

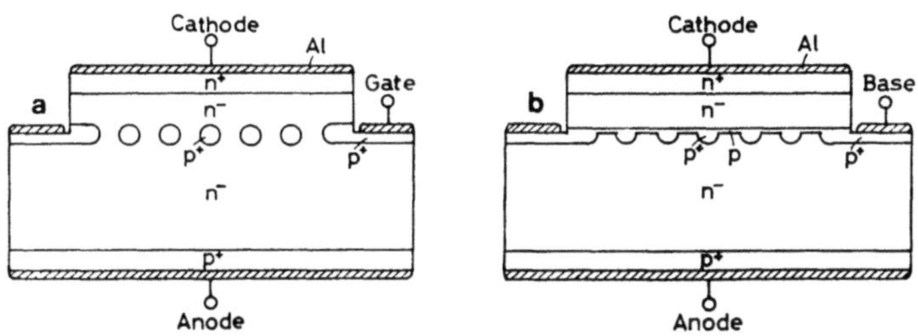

FIGURE 46. (a) Schematic cross section of static induction thyristor. (b) Schematic cross section of beam-base thyristor. (From Reference 44.)

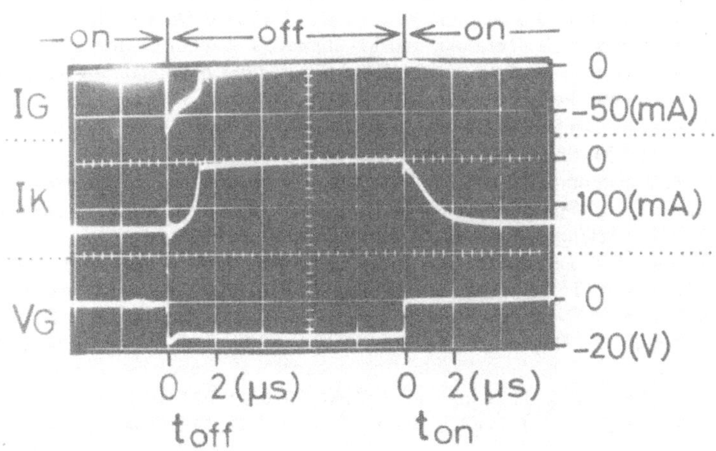

FIGURE 47. Switching characteristic of static induction thyristor. I_G is gate current, which flows only during turning-off. (From Reference 44.)

The SIT principle has also been extended to thyristors,[44] and has enabled very high-speed switching to be realized (<1 nsec). Two general kinds of device have been made; their structures are shown in Fig. 46. The beam-base device of Fig. 46b is intermediate in type between the static induction thyristor of Fig. 46a and the conventional *p–n–p–n* thyristor. One particularly interesting feature that static induction thyristors can offer is the possibility of dc switching, as shown in Fig. 47. This, of course, is not possible with conventional thyristors.

It is hoped that these examples confirm that the control of vapor epitaxial growth and the principle of lattice matching, both described in this chapter, do indeed have important practical consequences.

REFERENCES

1. T. U. M. S. Murthy, N. Miyamoto, M. Shimbo, and J. Nishizawa, *J. Crystal Growth* **33**, 1 (1976).
2. Y. Yasuda, K. Hirabayashi, and T. Morita, *Proceedings 5th International Conference on Solid State Devices*, Tokyo, 1973; published as a supplement in *J. Japan. Soc. Appl. Phys.* **43**, 400 (1973).
3. For example, on $SiCl_4$: H. C. Theuerer, *J. Electrochem. Soc.* **108**, 649 (1961); W. Steinmaier, *Philips Res. Rep.* **18**, 75 (1963). On $SiHCl_3$: J. H. Charig and B. A. Joyce, *J. Electrochem. Soc.* **109**, 857 (1962); A. M. Stein, *J. Electrochem. Soc.* **111**, 483 (1964).
4. M. Kumagawa, H. Sunami, T. Terasaki, and J. Nishizawa, *Japan. J. Appl. Phys.* **7**, 1332 (1968).
5. T. O. Sedgwick, J. E. Smith, Jr., R. Ghez, and M. E. Cowher, *J. Crystal Growth* **31**, 264 (1975).
6. J. Nishizawa, H. Nihira, T. U. M. S. Murthy, and M. Shimbo, to be published.
7. A. Masson, J. J. Metais, and R. Kern, *J. Crystal Growth* **3/4**, 196 (1968).
8. G. R. Booker and B. A. Joyce, *Phil. Mag.* **14**, 301 (1966).
9. M. Shimbo, J. Nishizawa, and T. Terasaki, *J. Crystal Growth* **23**, 267 (1974).
10. (a) J. Nishizawa, T. Terasaki, and M. Shimbo, *J. Crystal Growth* **13/14**, 297 (1972). (b) T. Shichiri, H. Kinoshita, and N. Kato, *Proceedings International Conference on Crystal Growth, 1966*, Pergamon, Oxford (1967) p. 385.
11. (a) J. Nishizawa and M. Shimbo, *J. Crystal Growth* **24/25**, 215 (1974); (b) J. Nishizawa, Y. Kato, and M. Shimbo, *J. Crystal Growth* **31**, 290 (1975).
12. J. Nishizawa and T. Terasaki, *Proceedings of International Conference on the Physics of Chemistry of Semiconductor Heterojunctions and Layer Structures (1970)*, Division of Technical Sciences of the Hungarian Academy of Sciences.
13. H. Sunami, T. Terasaki, N. Miyamoto, and J. Nishizawa, *J. Appl. Phys.* **40**, 4670 (1969).
14. J. Nishizawa, T. Terasaki, and M. Shimbo, *J. Crystal Growth* **17**, 241 (1972).
15. M. Shimbo, J. Nishizawa, and T. Terasaki, *J. Crystal Growth* **23**, 267 (1974).
16. H. Nihira, T. Shirasu, T. Terasaki, and J. Nishizawa, *J. Electrochem. Soc.* **122**, 781 (1975).
17. J. Roschen and C. G. Thornton, *J. Appl. Phys.* **29**, 923 (1958).
18. P. A. Totta and R. P. Sopher, *IBM J. Res. Dev.* **13**, 226 (1969); G. Ottaviani, S. Sigurd, V. Marrello, J. W. Mayer, and J. O. McCaldin, *J. Appl. Phys.* **45**, 1730 (1974).
19. Yao Ting Lee, N. Miyamoto, and J. Nishizawa, *J. Electrochem. Soc.* **122**, 530 (1975).

20. T. Abe, Y. Abe, and J. Chikawa, *Semiconductor Silicon*, H. R. Huff and R. R. Burgess (eds.), Electrochem Soc., Princeton, New Jersey (1973).
21. J. Nishizawa *et al.*, to be published.
22. Y. Watanabe, J. Nishizawa, H. Kon, T. Ikegami, and T. Matsushima, *Rec. Electr. Commun. Eng. Conv., Tohoku Univ.* **21**, 101 (1953).
23. T. H. Yeh, private communication (1961).
24. G. H. Schwuttke and H. J. Queisser, *J. Appl. Phys.* **33**, 1540 (1962); J. E. Lawrence, *J. Appl. Phys.* **32**, 1876 (1961).
25. J. Nishizawa and N. Miyamoto, to be published.
26. K. Yagi, N. Miyamoto, and J. Nishizawa, *Japan. J. Appl. Phys.* **9**, 246 (1970).
27. K. Gamo, A. Doi, K. Masuda, and S. Namba, *Proceedings 1st Conference on Solid State Devices*, Tokyo (1969), p. 78; published as a supplement to the *J. Japan. Soc. Appl. Phys.* **39**, 78–81 (1970). (Implanted nitrogen behaves as a donor at a substitutional site in silicon.)
28. F. Fujimoto, K. Komaki, K. Histake, and H. Nakayama, *Phys. Status Solidi (a)* **5**, 737 (1971); J. Haskell, E. Rimini, and J. Mayer, *J. Appl. Phys.* **43**, 3425 (1972).
29. R. O. Schwenker, E. S. Pan, and R. F. Lever, *J. Appl. Phys.* **42**, 3195 (1971).
30. L. Pauling, *Nature of the Chemical Bond*, 3rd ed., Cornell Univ., Ithaca, New York (1960).
31. J. Nishizawa, T. Terasaki, K. Yagi, and N. Miyamoto, *Proceedings Seoul International Conference of IEEE Korean Branch of IEEE, Seoul* (1970). See also Ref. 19.
32. N. Miyamoto, *National conference on crystal growth, Sendai 1976*, in *J. Japan. Assoc. Crystal Growth* **4**, 86–94 (1977). See also Ref. 19.
33. M. Watanabe, T. Yonezawa, M. Nakamura, T. Kato, M. Akatsuka, and T. Sakai, Dislocation-Free Diffusion, *Technical Report in the Technical Group on Semiconductor and Transistor JIEEC* (in Japanese) (1970).
34. F. Fujimoto, K. Komaki, M. Watanabe, and T. Yonezawa, *Appl. Phys. Lett.* **20**, 248 (1972).
35. G. Samsonov and V. Sleptsov, *Zhur. Neorg. Khim.* **8**, 2009 (1963).
36. Y. Watanabe and J. Nishizawa, Japanese Patent Issued No. 205068, No. of Publication 28-6055, applied for in November 1950; Y. Watanabe, J. Nishizawa, and Z. Yoshida, Conventional record of joint meeting of JIEE, JIEC, and JIEI No. 3.33 (1951), in: *Theory of Matter* (in Japanese) No. 41, 96 (1951).
37. W. Shockley, *Proc. IRE* **40**, 1289 (1952); W. Shockley and R. C. Prime, *Phys. Rev.* **90**, 753 (1953); G. C. Dacey, *Phys. Rev.* **90**, 759 (1953); W. Shockley, U.S. Patent Issued No. 279-0037, applied for in March 1952.
38. R. Zuleeg, *Solid State Electronics*, **10**, 449 (1967); R. Zuleeg and K. Lehovec, *Solid State Electronics* **13**, 1415 (1970).
39. S. Teszner and R. Gicquel, *Proc. IEEE* **52**, 1502 (1964).
40. S. Teszner, *IEEE Trans. Electron Devices* **ED-19**, 355 (1972).
41. J. Nishizawa, T. Terasaki, and J. Shibata, *IEEE Trans. Electron Devices* **ED-22**, 185 (1975).
42. J. Nishizawa and Y. Kato, *Proceedings of 6th European Microwave Conference, Italy* (1976); *IEEE Trans. on Electronic Devices* (March 1978).
43. J. Nishizawa and B. Wilamowski, *1976 International Conference on Solid State Devices*, A-2-6 in Tokyo.
44. J. Nishizawa and K. Nakamura, *Proceedings of 8th International Conference on Solid State Devices* (1976), C-2-3 in Tokyo, in *Japan. J. Appl. Phys.* **16**, 541–544 (1977).
45. T. Kodaka and J. Nishizawa, Report of the Research Institute of Electrical Communication, Tohoku University, Science Report RITU, B *Electrochem. Commun.* **22**, 63 (1971).

Chapter 3

The Verneuil Process

R. FALCKENBERG

1. INTRODUCTION

The idea and basic embodiment of the flame-fusion process were announced by Verneuil in 1902; initially the only purpose of the process was to make gem rubies.[1,2] The chief virtues of this method as far as today's technology and research are concerned are that it can be applied to a variety of oxides, and that no crucible is required to hold the melt. Problems connected with the use of a crucible, such as contamination by or reaction with the crucible material, are thereby avoided, an advantage that is more important the higher the melting point of the crystal.

In comparison with other methods not employing crucibles, such as float zone fusion, electron beam fusion, and skull fusion, the flame-fusion process has the advantage that it can be used for the growth of crystals from materials that do not conduct electricity. This advantage is restricted by the requirement that the crystal material be chemically compatible with the flame. The employment of a plasma torch is, in principle, a means of considerably increasing the number of substances that could be grown. However, as a result of the different flame geometry and temperature, it requires the process to be modified to such an extent that it will not be discussed here, and the reader must refer to the relevant literature.[3] The same can be said of arc-image devices.[4]

On account of the requirement for flame compatibility the Verneuil process is particularly attractive for high-melting oxides. It has consequently become an economic proposition to use this process for the manufacture of corundum and Mg–Al spinels, which have melting points between 2050 and 2150°C. In Central Europe alone there are half a dozen factories with certainly more than 5000 crystal-growing apparatuses, in

---

R. FALCKENBERG · Siemens Research Laboratories, Munich, West Germany.

which several hundred tons of these crystals are manufactured every year for use as gems and bearings.

In choosing a crystal-growing method, there are other factors to be considered besides the problem of growth potential. These are the desired and achievable crystal perfection and size, and the cost. As far as crystal perfection is concerned, the flame-fusion process is still inferior to other processes such as the Czochralski method (typical dislocation density $10^3\ cm^{-2}$).(5) State-of-the-art Verneuil crystals have a typical dislocation density of $10^5\ cm^{-2}$; subgrain formation can also occur. On the other hand, as the present chapter will show, there are ways of improving the crystal quality; poor crystal perfection is not considered inherent to the process.

With regard to the crystal size, it is now possible to grow sapphire crystals up to a diameter of around 75 mm.(6) Economic considerations play an important role when there is a great demand for crystals. The hydrogen consumption is a significant cost factor—a cheap source is a requisite for competitive ability. As an aid to the assessment of labor costs, one workman may operate about 50 apparatuses in present-day factories. A cost comparison on the basis of sapphire crystals with a large diameter (around 50 mm) shows that the Czochralski process is about twice as costly as the Verneuil process. The plant investment costs are in approximately the same ratio.

Despite the widespread employment of the process, there are very few papers that deal with the phenomenology(7) and the theory(8,9) of the process. The first reason for this is that in the temperature range of 2000°C and above, which is decisive here, there are data in only a very few instances on measured values and functional dependencies of such basic quantities as melt viscosity, interface (adhesion) and surface tension, thermal conductivity, and thermal capacity.(10–13) The other reason is that the high temperature makes it difficult for satisfactory measurements to be performed during growth.(14,15) In some areas this chapter will have to restrict itself to pointing out the problems connected with the subject, problems which, however, frequently only become evident after prolonged study of the process.

2. GENERAL DESCRIPTION OF THE PROCESS

In order to demonstrate the principle of the flame-fusion process, growth factors and the relevant parts of the apparatus are summarized in Fig. 1; examples of typical crystals and a simple apparatus as well as a scheme of the growth process are shown, respectively, in Figs. 2, 3a, and 3b. The starting material for the production of crystals is in the form of a

Growth Factors	Apparatus Factors
Powder	Powder container Feed mechanism
Flame	Burner
Melt film formation Growth Cooling	Growth chamber Pulling and rotation mechanism
↓ Crystal	

FIGURE 1. Growth and apparatus factors for the Verneuil process.

fine-grained powder. The top of a seed crystal is heated by an oxy-hydrogen flame until a thin melt film develops. The powder is now scattered from a supply container into the flame and reaches the melt film in a more or less molten form. An amount of matter corresponding to the powder flow crystallizes continuously out of the film; the crystal is pulled downward at the same rate as it is growing toward the flame. Heat losses, which are essentially due to radiation, are diminished by enclosing the crystal in a growth chamber. When the desired crystal length is reached, powder flow and crystal motion are stopped and the flame extinguished.

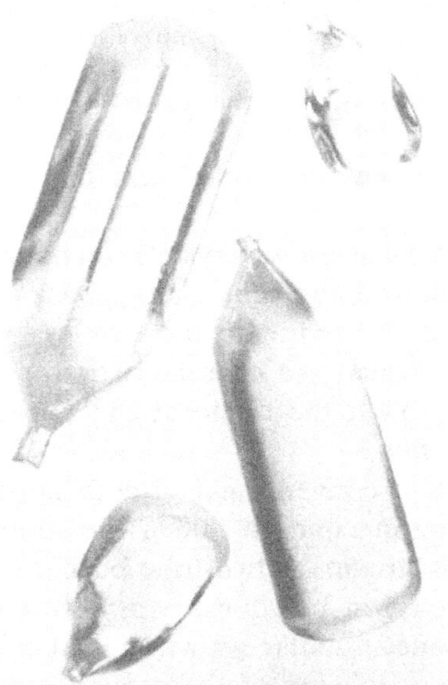

FIGURE 2. Mg–Al spinel single crystals grown by the flame-fusion process.

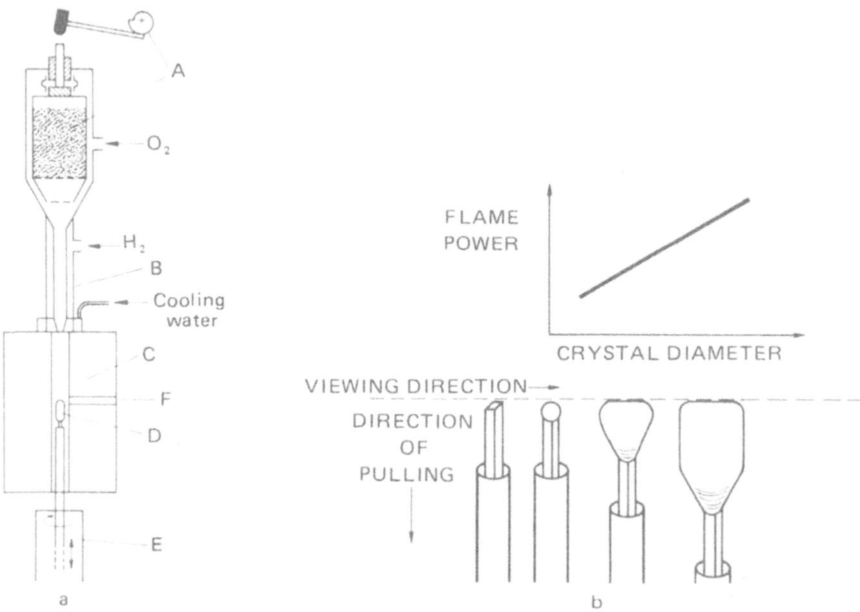

FIGURE 3. (a) Conventional Verneuil growth apparatus: A, Powder container and feed mechanism; B, burner; C, insulated growth chamber; D, crystal; E, vertical movement and rotation mechanism; F, observation port. (b) Scheme of the growth process. By melting the top of the seed, a "melt pearl" is formed which is covered by a thin melt film. The crystal is pulled downward and powder is fed in such a way that the growth front stays at a constant level. Broadening is done by increasing the flame power as shown in the diagram. If there is no further increase in power, the crystal grows with constant diameter. Generally, the relation between flame power and crystal diameter is not as straightforward as shown but depends additionally on the interactions between flame, crystal, and furnace.

After being cooled down the crystal can be taken out of the growth chamber.

The shape of typical as-grown crystals can be seen from Fig. 2. Usually the seed crystal is followed by a broadening phase which leads on to a main part of constant diameter. Frequently more or less pronounced planes form along the side faces, which are characteristic for the crystal and growth conditions. During growth the dome at the upper end of the crystal is covered by the melt film.

Figure 3a shows a conventional growth apparatus. The powder is scattered by a tapping mechanism out of the supply container through a sieve into the oxygen stream. A two-tube burner is used to add hydrogen concentrically to the oxygen–powder stream. Flame and crystal are enclosed in a vertically positioned tubular growth chamber.

Figure 3b shows a scheme of the growth process with the development of a crystal out of an initially formed small melt pearl. The influence of the flame power on the crystal diameter is indicated.

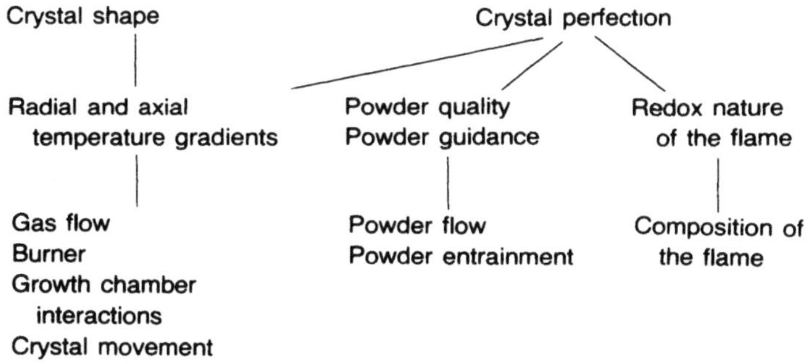

FIGURE 4. Relations between crystal properties, their physical and chemical causes, and process factors.

In principle, this process is simple; however, it turns out to be complex as soon as the relations are considered between crystal properties and process factors. These are outlined in the block diagram in Fig. 4, which connects crystal and process factors and relevant physical and chemical quantities. The crystal can be characterized in terms of its shape and perfection. As can be seen by considering the processes of growth and cooling, the axial and radial temperature gradients in the crystal are of basic importance for both these properties. The gradients are defined by a series of process factors. Variables are gas flow and crystal motion; relatively fixed parameters are the geometry of the burner and of the growth chamber as well as the pyrometric design of the growth chamber. In addition, interactions between flame gas, growth chamber, and crystal have to be considered.

In order to adjust the temperature gradients in the crystal it is necessary to bring heat onto the crystal in a directed manner. As far as the melt face is concerned, the radial temperature distribution can largely be adjusted by suitable burner construction and gas flow. Heating of the side faces of the crystal, however, presents a major problem of the process. Flame power can be transferred in a quantitative manner only by direct impact onto an object, but not by grazing incidence. That means that the side faces of the crystal cannot be affected directly by the flame. This is made clear by Fig. 5, which shows the gas flow in a conventional tubular furnace. The flame gases hit the crystal top directly, lose a large part of their energy, and are deflected laterally. If the distance to the inner wall of the growth chamber is small, the horizontal component of the flame gas adds to the temperature rise of the wall. Essentially it is only via the radiation interaction with the inner walls of the growth chamber that the side walls of the crystal can be heated or, more exactly, its radiation losses reduced. Schematically the effect of the burner and the interaction between crystal and growth chamber is presented in Fig. 6.

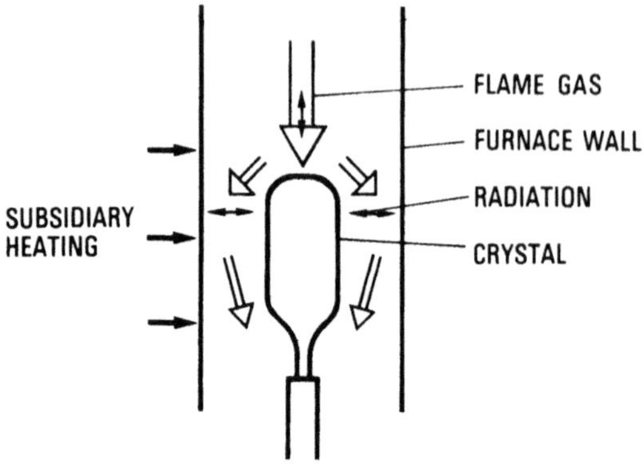

FIGURE 5. Interaction between flame gases, growth chamber, and crystal. There is also interaction by radiation between the crystal on one hand and the growth chamber and burner on the other. The action of subsidiary heating is indicated.

The thickness of the arrows indicates the intensity of the energy transfer. Quantitatively the transfer of heat by convection is estimated to be about 10% of that transferred by radiation from wall to wall.[8] If the interaction is not sufficient to obtain the gradients required, auxiliary means have to be applied such as improvement of chamber insulation, additional lateral heating, and changes of the geometry of the flame gas flow. Radiation loss from the crystal top to the burner can be controlled by adjusting the reflectivity of the lower side of the burner.

The interaction between growth chamber and crystal is time dependent until thermal equilibrium has been established, i.e., essentially until constant crystal diameter has been reached. Radiation density and transfer of the flame power increase as the distance of the crystal to the chamber wall decreases; in addition there is the influence of the heat

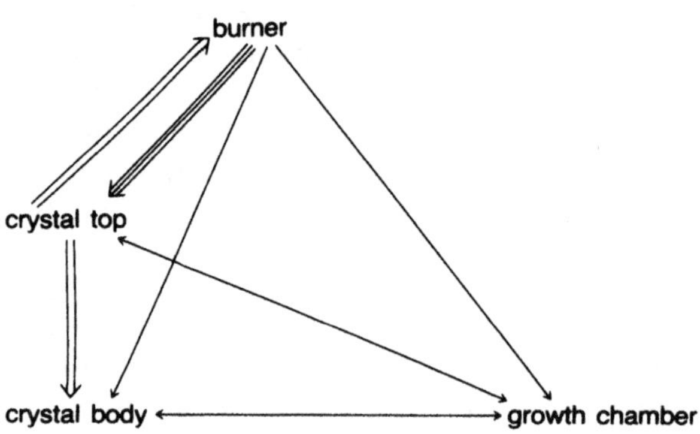

FIGURE 6. Thermal transfer during growth schematically depicted. The side faces of the crystal are heated essentially only by radiative interaction with the walls of the growth chamber and experience little direct interaction with the main flame.

capacity and thermal conductivity of the growth chamber material, which causes a time-dependent increase of temperature in the growth chamber.

With regard to crystal perfection, Fig. 4 shows that it also depends both on flame properties and on the powder supply. The transition from powder particle to liquid droplet requires small grain size, low density, and fairly constant powder flow; a further requirement is chemical purity. Parameters are the technique of powder feeding and the method of powder production; a variable is the powder flow rate. In addition the redox nature of the flame may be of importance if components of the material to be grown can occur in different valence states.

If the parameters burner, growth chamber, powder quality and feed, and flame composition are all set, the growth procedure is defined by the variables. Generally these are adjusted in the following way: Constant growth rate is postulated. The pulling rate is therefore kept constant and the powder flow is controlled in such a way that the position of the crystal top, i.e., of the growth front, remains unchanged. The value of the growth rate is determined in practice by the fact that at too high a rate the crystal perfection, especially the mechanical strength, decreases. Theoretical considerations for sapphire which take the removal of the latent heat as the limiting factor lead to much larger values.[8]

Rotation, an additional possibility of crystal motion, is of marked advantage if crystals of large diameter, say above 40 mm, are grown. In this way unsymmetrical growth can frequently be prevented. Fast rotation, above 135 rpm, has been found to influence the form of the growth front, since the melt is driven outward by centrifugal force.[16]

The gas flow controls the broadening of a crystal; at constant crystal diameter only the radiation losses have to be compensated for, and these increase as the crystal surface increases. With sapphire and Mg–Al spinel, for example, crystals are grown in an excess-hydrogen atmosphere; therefore, a change of power can be brought about by simply changing the oxygen gas flow.

After this short description of the effects of the process factors let us return to Fig. 1. The starting point in the growth procedure is the powder. The small thickness of the melt film (discussed in the following section) presents a danger for the perfection of the crystal as long as the powder reaching the film is not entirely molten. In addition the broadening out of a crystal occurs as a consequence of a series of controlled melt runovers which necessitates close control of the melt film. Finally in the conventional Verneuil process the cooling of the grown crystal is critical because of the difficulty of controlling the temperature gradients. Stresses develop which lead to plastic deformation of the crystal, i.e., to the occurrence of slip and twinning, or even to fracture. Residual stresses may finally be removed by annealing.

3. THE MELT FILM

3.1. Thickness and Shape

The melt film thickness was determined by Grabmaier[17] from the profile of the chromium concentration when ruby was grown on sapphire. The method is universally applicable in that most crystals can be fed a dopant, whose change in concentration at the interface to the undoped crystal can be measured over distances on the order of a few micrometers with a measuring instrument such as a microprobe.

For a concentration c the amount of material in the melt film is $q\delta c$, where q is the cross-sectional area and δ is the average thickness of the film. During growth an amount of material equal to that added to the melt film crystallizes out on the underside of the film, the film thickness remaining constant. On adding undoped material, dopant depletion takes place in the melt film. If possible, evaporation is also taken into account; the impurity concentration as a function of the space coordinate x and the time t may be represented by $c(x, t)$. If the crystal grows by a thickness dx through the addition of undoped material, the reduction in the impurity level in the melt film is represented by

$$-q\delta\, dc(c, t) = kqc(t, x)\, dx + \alpha qc(c, t)\, dt$$

k being the distribution coefficient of the impurity in the matrix and α the evaporation coefficient of the impurity. The profile of the impurity concentration in the doped/undoped matrix transition area is obtained by integration:

$$\ln c = \ln c_0 - \frac{k}{\delta} + \frac{\alpha}{\delta v} x$$

where $v = x/t$ is the growth rate of the crystal. In the case of the ruby–sapphire interface the evaporation term could be neglected; the thickness of the melt film was found to be around 20 μm.

Adamski *et al.*[18] also obtained a value of approximately 20 μm by simply measuring the boundaries between the white and red parts of the sapphire–ruby interface with a microprobe; using this method a value of about 40 μm would be obtained from the curves measured by Grabmaier.

At variance with these values, a melt film thickness of 150 to 200 μm was measured by Shiroki in the growth of ruby.[15] An explanation of this discrepancy could be that Shiroki did not perform his measurements during the actual growth process. Following machining, fully grown crystals were placed in the furnace again, the top was melted, and the melt film thickness was marked with Cr_2O_3 after a steady temperature

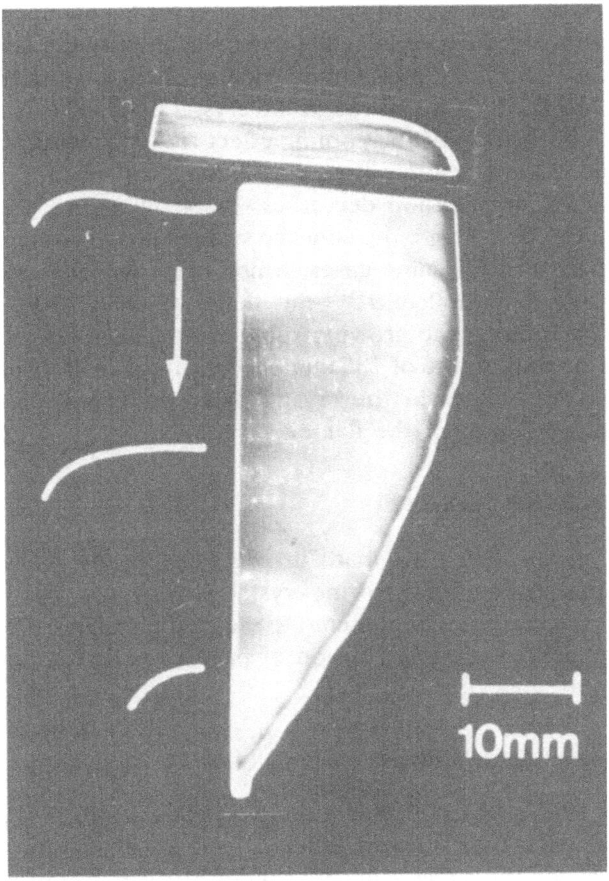

FIGURE 7. Autoradiograph of aluminum distribution in a Mg–Al spinel single crystal. This reveals traces of the growth front at different stages of the growth process.

distribution had been achieved. It is not certain whether the melt film thickness existing during the actual growth process is reproduced in this way.

It should be noted that melt film thickness is not always the same even with apparently similar crystals. It depends on a number of parameters which influence the temperature distribution in and around the melt film, such as growth rate, length of crystal grown, and the material which is being grown, as well as on burner and furnace design. For large-diameter crystals it also varies with the distance from the growth axis.

The outlines of Al-rich striations in Mg–Al spinels (Fig. 7) show how the shape of the melt film changes, as the crystal cross section increases during growth, from convex via flat to concave; only the shoulder at the edge remains essentially unaltered.[19] These changes in shape are partly determined by the heat lost through the crystal body. This is initially large due to the penetration of the cooling effect from the conical crystal foot, where its downward heat-emitting area is large in relation to the crystal volume. As this penetration decreases, the growing areas near the axis flatten out. The final stage, the concave surface, may be due in part to the pressure of the central flame gases, which press the melt outward.

According to Kvapil *et al.*[16] the shape of the growth front can be influenced by rotating the growing ruby crystal. At a crystal diameter of 20 mm and a rotation rate of 135 min^{-1} a concave growth front developed, which in an area close to the crystal axis approximately matched the shape of the isotherms of the flame.

3.2. Effects in the Melt Film

Some effects which originate in processes in the melt film are the cross-sectional shape of the growing crystal, growth striations with diffusion movements (separation phenomena), penetration of heat sinks, the shape and arrangement of bubbles, fusion of powder particles, and convective flow.

3.2.1. Cross-Sectional Shape

During the growth of, for example, Mg–Al spinel crystals, which have a cubic structure, various kinds of polygonal crystal cross sections are observed for the same seed orientation. In the case of growth in the ⟨100⟩ direction, for example, rectangular and octagonal crystal cross sections can occur without the cylindrical symmetry of the growth apparatus being altered. According to Neuhaus and Brenner,[20,21] the rate of heat loss through the crystal and the temperature gradient at the melt film–crystal phase boundary of the crystal rim are the determining factors. If the heat loss, and therefore the temperature gradient, is low, the cooling effect only penetrates slightly beyond the crystal–liquid phase boundary into the adhering melt film, and thus gives the molecular attachment energies varying with crystal orientation only little chance of taking effect. The melting point isotherm, circularly symmetrical by nature, thus essentially determines the appearance of the growing single crystal—the growth in round fashion is approximated by the octahedral cross-sectional shape. If heat dissipation and radial temperature gradient are greater, the penetration of the cooling effect through the crystal–melt film phase boundary into the latter is greater, so that a considerably wider zone of undercooled metastable melt is maintained between the growth front and the melting point isotherm. In this zone the varying molecular attachment energies come into effect to a greater extent, so that the result is square cross-sectional growth with only the extreme edges rounded off. In actual fact it has been found by experiment that square crystals were obtained in relatively poorly insulated furnaces, octagonal crystals in well-insulated furnaces (see Fig. 34).[7] The influence of the Verneuil apparatus regime on the growth shape of α-corundum crystals was studied by Akimovich.[22] Changes in the growth rate along a geometrical axis within the limits of 2 to 12 mm/min and in rotation velocity around the same axis within the limits of 2 to 20 rotations/min are found not to affect the outer shape of a growing crystal. It is possible under some thermal conditions to obtain α-corundum crystals the shapes of which are very close to the equilibrium growth shape. These conditions are determined by the percentage composition of the hydrogen–oxygen mixture and the rate of its consumption. Moreover, in this case there appear $\{0001\}$, $\{10\bar{1}1\}$, $\{11\bar{2}0\}$, $\{10\bar{1}0\}$ faces of equilibrium growth shape.

According to Zemlicka,[23] the cross-sectional shape of sapphire crystals of large diameter depends on the orientation of the growth axis. An angle of 52 to 55° between the *c* axis and the growth axis is favorable; with a larger angle, the crystal shape becomes irregular; the formation of the basal plane $\{0001\}$ is especially favored and could not be suppressed. Bagdasarov *et al.*[24] investigated the anisotropy of the thermochemical dissolution rates of sapphire crystals. The studies permitted conclusions to be drawn about the growth rates along different crystallographic directions. Sapphire crystals will grow along the direction of the normal to the face $\{11\bar{2}3\}$ at the maximum rate and along the $\langle 11\bar{2}0\rangle$ and $\langle 0001\rangle$ directions at the minimum rate. As is well known, crystals exhibit slow growth faces; the direction of the least growth rate is also the direction of the least dissolution rate.

3.2.2. Striations

Periodic striations parallel to the growth front have often been observed in crystals. These are just as likely to occur in crystals grown from a solution as in crystals manufactured using the Czochralski or float zone method. The reason for the occurrence of a striation is always a periodically varying growth rate; the variations of the growth rate may be due to apparatus faults (e.g., play in lead screw or bearings), temperature fluctuations, or mechanism of crystallization.[25,26] Striations have been observed in Verneuil crystals[19,27,28]; an example with Al_2O_3-rich Mg–Al spinel has already been mentioned (Fig. 7). Because of the pronounced two-phase nature of this crystal, which had the composition $MgO:Al_2O_3 = 1:2$, striations can be particularly well observed. The spacing between the striations is 0.12 cm and the period is 4.8 min (pulling speed 15 mm/hr). On careful examination a superimposed finer striation can be detected with 0.03-cm spacing and 1.2-min period. The powder was supplied via a feeder sieve, vibrated at a frequency of 100 Hz; any relationship between the feeder frequency and the frequency of the striations can therefore be ruled out. (See also Section 6.4.)

The chromium distribution in a ruby crystal which was manufactured using the tapping method for delivering powder, was investigated by Pink.[27] Owing to the small crystal diameter the striations are convex along the entire length of the crystal. It was observed that the frequency of the striations was a factor of 5 lower than the frequency of the tapping mechanism.[17]

An effect of the periodic segregation entering into the visible range is shown in Fig. 8. It is the periodic formation of bubble rows (the reasons for the formation of bubbles are dealt with in Section 3.2.3). The composition of this spinel crystal is $MgO:Al_2O_3 = 1:3.5$; the spacing and period of the striations are 0.01 cm and 30 sec, respectively. The small

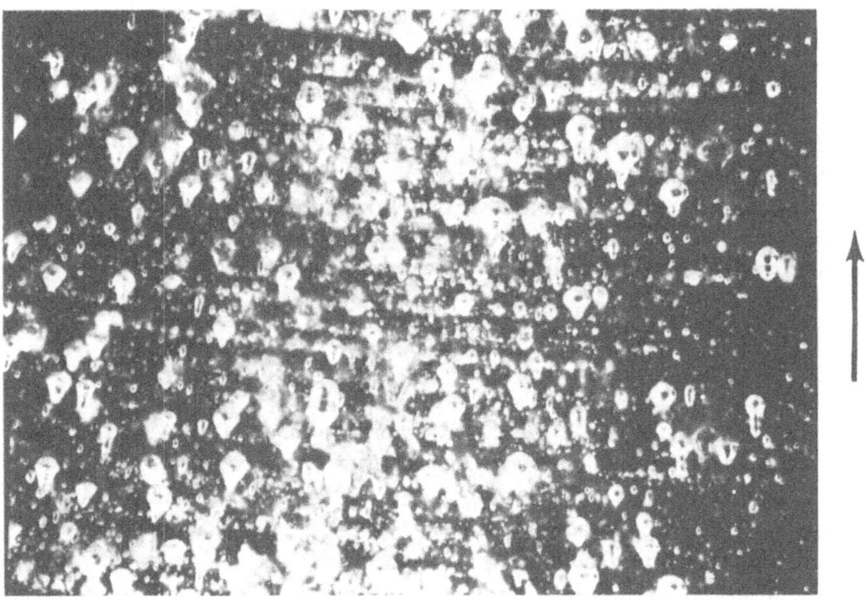

FIGURE 8. Periodic ordering of bubbles in an Mg–Al spinel single crystal of the composition $MgO:Al_2O_3 = 1:3.5$. The arrow shows the direction of growth. ×22.

bubbles are expelled during the phase with slow freezing rate, while they are trapped in the phase of rapid freezing (see the equation in Section 3.2.3). The large bubbles are mostly based in the striae formed by the small bubbles and are pear shaped, the narrow part always facing away from the direction of growth. One can thus conclude that large bubbles initially protrude a certain distance out of the melt film. As the melt film surface rises the bubbles are first compressed; the internal pressure increases and causes the upper end of the bubble to attain a spherical shape as the melt surface rises further. According to these considerations the thickness of the melt film should be smaller than the distance between two neighboring rows of bubbles, but larger than the diameter of the small round bubbles: $40 < \delta < 100\ \mu m$ for Mg–Al spinel of composition $MgO : Al_2O_3 = 1:3.5$ in an area close to the crystal axis for the growth conditions mentioned.

The origin of the different kinds of striations mentioned has not been investigated. In addition to the reasons already mentioned, however, cyclical variations in gas pressure have to be considered, arising, for example, from pressure-reducing valves which are diaphragm-operated.

3.2.3. Bubbles

If the growth conditions are unfavorable, bubbles can be formed during the crystallization process. A distinction can be made between irregularly distributed single bubbles or bubble groups and a bubble pipe occurring along the crystal axis. In both cases the reason for their formation is generally that powder particles do not fuse so fully in the flame that all the gas can escape from the powder. The powder is in any case very voluminous, particularly if it is manufactured by the thermal decomposition of alum. For such sapphire powder the bulk density is $0.2\ g/cm^3$, while the specific gravity of the sapphire crystal is $3.96\ g/cm^3$.

There is thus a considerable reduction in volume caused by the air being expelled from the existing cavities. If decomposition of the alum is faulty, there may also be ammonia or sulfate residues in the powder; certain residues of trapped water at about the 1% level are present in any case. All these volatile constituents must be expelled as the powder passes through the flame, or at very least in the melt film, if a transparent crystal is to be obtained.

Figure 9 shows a crystal inside of which a "pipe" of bubbles has been formed. This type of bubble formation can be observed if too little energy is transferred onto the powder on its way from the burner nozzle to the crystal. The energy transfer is defined by a number of apparatus parameters which characterize flame size, density of flame gases, and

FIGURE 9. Bubble pipe in a Mg–Al spinel crystal of composition $MgO:Al_2O_3 = 1:2$. (×1.3.)

surface temperature of the flame gas pipe in the furnace. The parameters particularly involved are the diameters of the nozzle D_n and crystal D_c, distance L_f between nozzle and growth front of the crystal, as well as heat conductivity λ and diameter of the flame gas pipe D_f. Experimentally one finds that bubble formation sets in if $A > K$, where A is a term that increases with D_n, D_c, D_f, λ, v_p, and decreases with L_f. K is an apparatus constant; v_p is the pull rate which accounts for the time a particle stays in the melt film. This dependence was observed using a two-tube burner; with multitube burners (K_M) the bubble formation seems to be less critical, i.e., $K_M > K_2$. Overheating of a crystal may also cause the formation of bubbles which partly migrate out and subsequently burst. This effect can easily be distinguished from the one just mentioned since it occurs even without powder being fed.

The occurrence of single bubbles or bubble groups[29] is little influenced by altering the apparatus so as to reduce A. There are usually involved accumulations of powder which have formed somewhere between sieve and crystal top and free themselves when their weight becomes too large. Such powder "avalanches" cannot be fully fused because of their large size. It is therefore necessary to eliminate the causes of the powder accumulations. The burner tube through which the powder stream falls should be well polished and the parts through which the powder is transferred from the shaking or vibration funnel to the burner tube should be examined to see whether powder accumulations can occur. If a vibrator is used, the burner tube usually vibrates enough to prevent the powder piling up; Shankland(30) recommended a special solenoid which acts on the funnel and prevents such accumulations of powder.

3.2.4. Temperature Conditions in the Vicinity of the Melt Zone

The melt–crystal interface coincides—when the undercooling previously mentioned is ignored—with that of the melting point isotherm; it is therefore an area of constant temperature. In a region in the vicinity of the growth front the isotherms in the crystal run parallel to the solid–liquid interface. As a result, the radial temperature gradient is practically zero in areas near the axis or increases a little toward the outside in very broad crystals (compare with the shape of the striations in Fig. 7); in a narrow zone near the edge the temperature rises steeply as the radius increases, corresponding to the melt film moving downward from the crystal top. The almost constant radial temperature characteristic inside the crystal was confirmed by Shiroki(15) through measurements made on ruby, which cover the region from 2 to 6 mm below the liquid–solid interface. Unfortunately no measurements have been made in the zone near the edge, approximately 2.5 mm wide, where according to photoelastic examinations, great stresses can occur,(31–33) and where facet effects may be most apparent.

Nothing is known of the temperature distribution in the melt film itself. From the manner in which the thermal energy is applied, all that can be said is that the surface of the melt is the hottest part. Thermal convection in the melt is therefore ruled out (there is still, however, the possibility of solutal convection induced by segregation at the solid–liquid interface). A temperature jump is to be expected at the liquid–solid interface due to the reflection of radiant heat (at the refractive index step), since at the prevailing high temperatures the heat transport due to radiant heat greatly exceeds that due to thermal conductivity.

An impression of the interaction between the melt film and the neighboring crystal zone can be obtained by reference to Fig. 6. It should

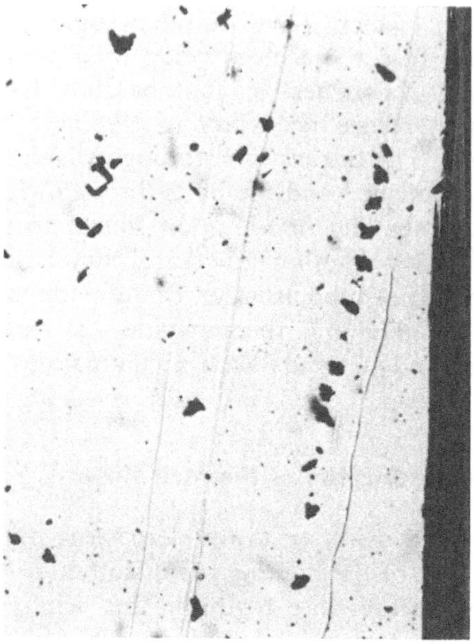

FIGURE 10. Rows of incompletely melted powder particles aligned parallel with the growth axis (Mg–Al spinel crystal, ×50).

be noted that the heat of fusion (26 kcal/mol for sapphire) can be neglected because it is <1% of the heat transferred to the melt surface by the flame compensating the heat loss by radiation.

The axially parallel lineup of unfused powder particles (Fig. 10) occasionally observed when the powder supply is coarse-grained and the pulling rate high is attributed to incompletely fused particles entrapped in the crystal acting in effect as a heat sink for a certain period. The melt film is particularly thin over such locations and larger particles newly arriving there become less likely to be completely fused.[19]

4. PHENOMENOLOGY OF THE GROWTH PROCESS

The growth of a crystal with constant cross section is usually preceded by a process of broadening of the seed crystal. Assuming the previously mentioned conditions (Section 2) of constant growth rate, a powder flow sufficient to keep the growth front at constant height, and excess H_2 gas flow, crystal broadening is obtained by increasing the O_2 gas flow ϕ. Broadening means a continuous controlled rundown of melt toward the rim of the crystal top, where the melt solidifies. If b denotes the change of the crystal diameter perpendicular to the longitudinal axis per unit of time, the broadening angle α (Fig. 11) is obtained from

$$\tan\frac{\alpha}{2} = \frac{b}{v}$$

v being the pulling rate. In the following some visual observations are reported, a model accounting for the broadening process is described, and starting points for a theory are discussed.

4.1. Visual Observations

With Mg–Al spinel and sapphire it is possible to recognize visually the liquid–solid boundary line at the top of a crystal during growth. By observing the boundary line some idea can be obtained about the mechanisms involved in the formation of a crystal by the flame-fusion method.

Growth begins with the formation of the so-called melt pearl. The melt pearl (Fig. 12) develops without the addition of powder from the substance of the seed crystal when the oxygen flow rate ϕ is increased. Formation of the melt pearl is not such that the seed crystal melts from the top to the lower end of the melt pearl, but material first melts away from the upper four corners of the rectangular seed rod, later from the whole top, and then is deposited again in layers on the sides. During this process the bulk of the melt pearl remains solid except for a very thin surface layer, with a thickness not exceeding some 60 μm at the

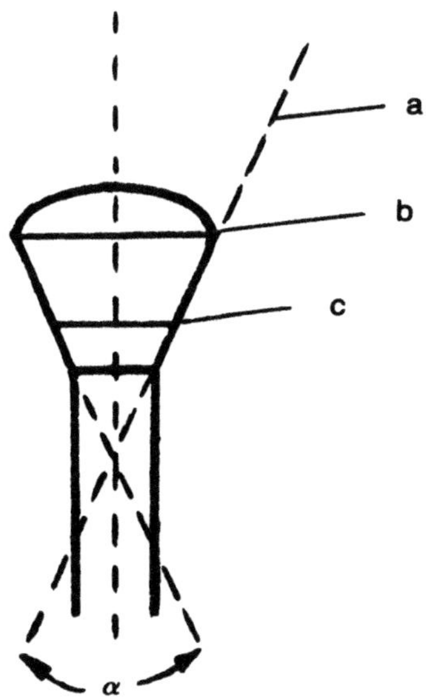

FIGURE 11. Initial broadening of a growing sapphire crystal. a, Direction of growth along which broadening occurs; b, line where momentary protrusion develops; c, liquid–solid boundary line where melt film ends.

uppermost point in the case of Mg–Al spinel. This can easily be seen by examining between crossed polarizers a vertical cut through the solidified melt pearl. The pearl shown in Fig. 12 was cooled down by quenching the flame; as can be seen from the four ⟨111⟩ faces it solidified in single-crystal form. It should be noted that the melt pearls are frequently uniformly round and that the molten film always extends over the spherical part down to the seed crystal. The circular shape at right angles to the growth axis reflects the cylindrical symmetry of the flame (Fig. 12b). For completeness it may be mentioned that crystals of random orientation can be grown from a sintered powder cone, the tip of which is used to produce a melt pearl.

With the melt pearl formed the powder feeder is now turned on and simultaneously one begins to draw the crystal downward. When the oxygen flow is increased, broadening starts. The liquid–solid boundary line initially stays unchanged at the seed crystal until a definite conical crystal form has developed. The boundary line then moves slowly upward until, at a crystal length of 10 to 15 mm, depending on the parameters of the apparatus, it coincides with the line describing the maximum circumference of the growing crystal. From visual observation it seems to be certain that during this initial phase—at small α—the crystal only increases in size directly at or directly above (and not below) the maximum diameter, even though a molten film covers the crystal below

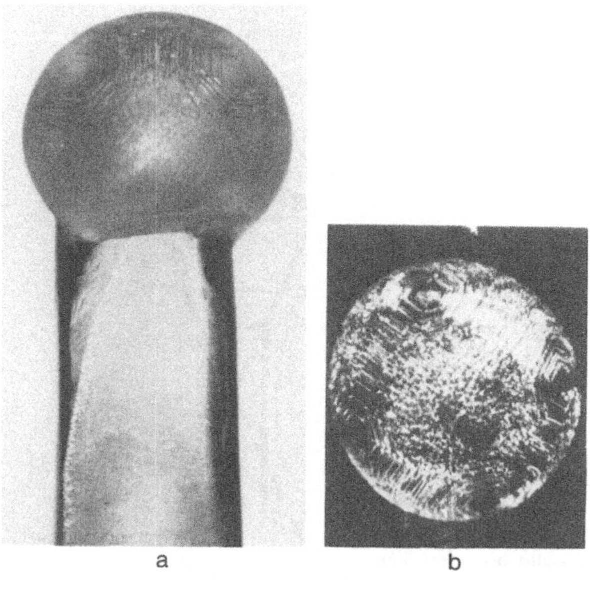
a b

FIGURE 12. (a) Melt pearl of Mg–Al spinel made from a crystal seed without adding powder. Crystallization was induced by suddenly extinguishing the flame. Only after the melt pearl has been formed does the growth process begin. (b) Top view of the melt pearl shown in (a). Both photographs ×7.

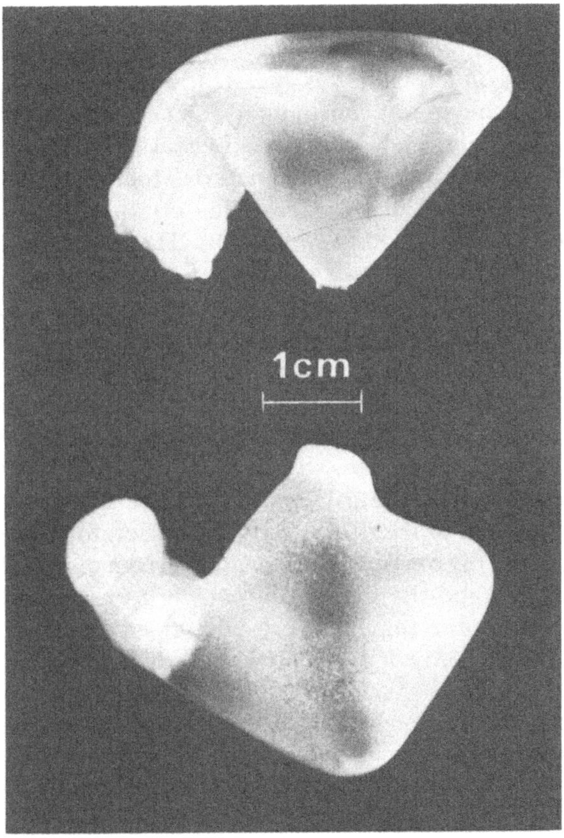

FIGURE 13. Melt runover. As a result of excessive oxygen increase, the melt volume has increased to such an extent that surface tension is no longer adequate to keep the melt from running down the side of the crystal. The bottom photograph shows a view of the same specimen from above.

this line (Fig. 11). Further broadening of the crystal has to be done at an angle $\alpha \leqslant \alpha_{max}$ by adjusting the O_2 gas flow ϕ (Fig. 15). α_{max} is defined by crystal and furnace[7]; if $\alpha > \alpha_{max}$, the melt drops down (Fig. 13).

Finally, steady state conditions will be reached. The crystal grows with constant cross section at constant gas flow ϕ. An important observation in this context has been made with Mg–Al spinel crystals quenched out of the steady state growth phase: Longitudinal sections show that the thickness of the melt film approaches zero as the melt film approaches the liquid–solid boundary line.

4.2. Model of Broadening

The initial point (Fig. 14) is the steady state, i.e., the crystal grows with constant diameter d at constant oxygen flow ϕ (and with much excess of hydrogen) (see Fig. 14a). When ϕ is increased by a small amount $\Delta\phi$ a thin, already crystallized, layer of material melts again (Fig. 14b). The excess melt flows down at the rim of the crystal top to the liquid–solid boundary line where it forms a protrusion. The position of the surface of the liquid band which immediately borders the boundary line is defined by the angle θ (Fig. 14c). θ is formed by the surface normal and a line perpendicular to the growth direction. As θ increases, the thermal loss caused by radiation from the protrusion increases and finally leads to solidification when $\theta = \theta_s$ (Fig. 14d). The process continues as long as excess melt is being formed by the increased oxygen flow. The crystal subsequently continues to grow with constant diameter $d + \Delta d$ (Fig. 14e). If the oxygen flow rate is again increased by $\Delta\phi$, the process is repeated. The opening angle α of the growing crystal is therefore determined by the rate of increase of oxygen flow or, in practice, by the amount of additional oxygen and the intervals at which it is supplied (Fig. 15).[7] In the growth context the remelting crystal matter is replaced by melt that has not solidified, provided by powder fed to the crystal during the broadening.

The temperature conditions near the line where the protrusion is formed are decisive for broadening. On the one hand, it must be possible for melt to flow and collect up without solidifying at the upper outer

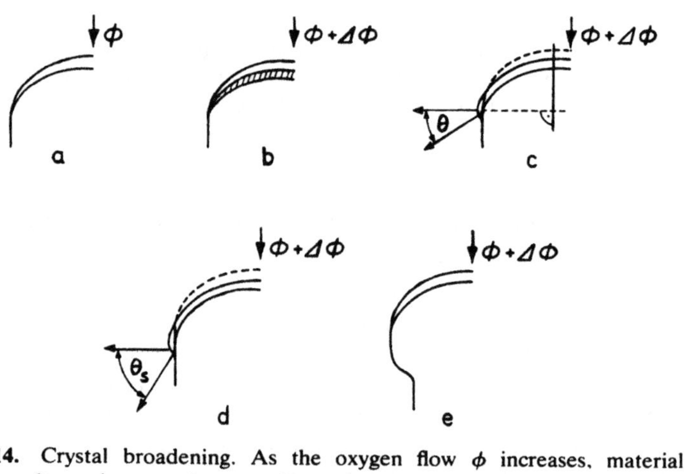

FIGURE 14. Crystal broadening. As the oxygen flow ϕ increases, material already crystallized melts again, runs down to the crystal rim, and forms a protrusion—the lower side of which crystallizes.

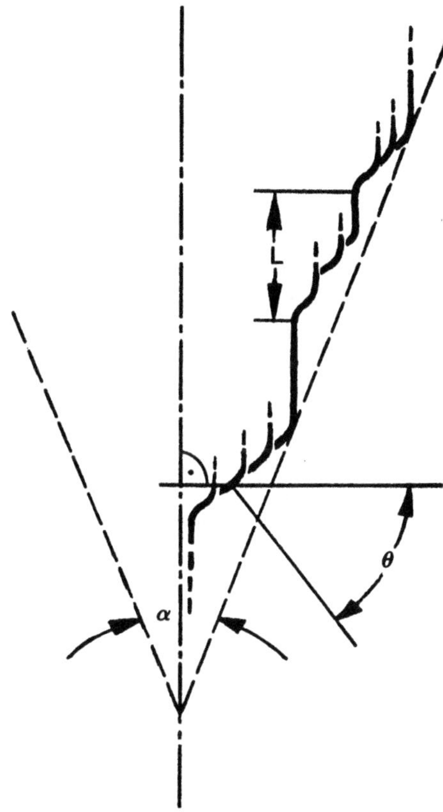

Figure 15. Broadening angle α of a crystal as the oxygen flow ϕ is increased stepwise.

crystal zone, while, on the other hand, it must be possible for the underside of the protrusion to solidify; i.e., there must be a suitable temperature decrease at the lower outer crystal zone. The main factors in determining these conditions are the flame shape and the thermal properties of those parts of the walls of the growth chamber for which there is an interchange of radiant heat with these crystal zones. In practice it is important that there must be a flame component impacting onto the upper edge zone of the crystal while the radial temperature gradient must be small at the level of the liquid–solid transition line. (The realization of these requirements necessitates suitable development of burner design and growth chamber construction.) If in the course of the broadening process these requirements are no longer fulfilled, for example because the diameter of the burner gradually becomes small compared with the diameter of the crystal, the following effect is observed: Further increase in the oxygen gas flow no longer leads to the formation of a protrusion, but to an increase of the melt volume in the inner part of the growth front only. A dish-shaped melt volume forms. With still further increase in the O_2 gas flow the rim melts at some thermally favored spot and the melt runs over (Fig. 34d).

4.3. Starting Points for a Theory

The physics of the broadening process, i.e., of the controlled rundown of the melt, are complex. At steady state conditions, the melt film thickness approaches zero at the solid–liquid boundary line. With sufficiently thin liquid layers the interface tension between liquid and solid is large enough to keep the melt from dropping down. As the O_2 gas flow increases by $\Delta\phi$, excess melt runs toward the boundary line. As long as $\Delta\phi/\Delta t$ stays below a certain small value A, $(\Delta\phi/\Delta t) < A$, the interface tension can keep the melt from dropping. A becomes large for a high solidification rate at the crystal rim and a low flow rate of the liquid, i.e., high viscosity of the melt and a low density. As $\Delta\phi/\Delta t$ increases $[(\Delta\phi/\Delta t) > A]$, surface tension becomes effective in addition to the interface tension. This section is limited by an upper value B $[A < (\Delta\phi/\Delta t) < B]$ above which surface or interface tension is no longer sufficient to retain the melt. For $(\Delta\phi/\Delta t) > B$, the melt drops down. The dependence of the broadening angle α on the change of the gas flow $\Delta\phi/\Delta t$ is indicated in Fig. 16.

In a theory of the process, surface and interface tensions and the transport mechanism in the melt would have to be considered. The transport mechanism would include considerations of the heat balance; the melt flow would be formulated by a melt source and a melt sink characterized by a viscosity $\eta(T)$ and a certain solidification rate. The shape of the crystal top is defined by gas flow dynamics,[9,34] and by the heat transfer from gas and furnace to the crystal top. Several other factors are considered to be quantitatively less important, such as change of heat conductivity, specific heat, and redox potential of the gas. The redox potential determines the content of suboxides in the gas; the suboxides

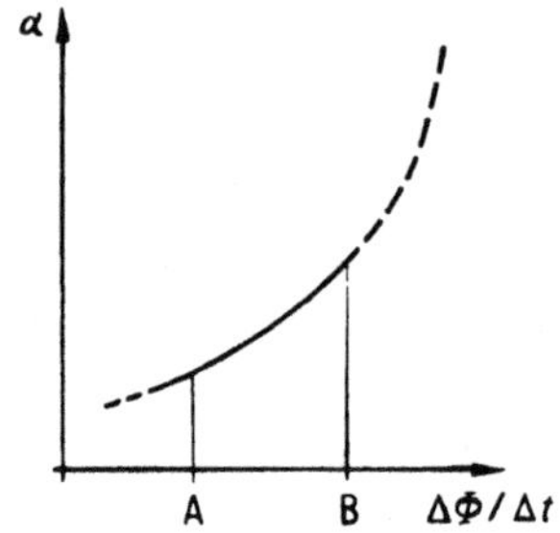

FIGURE 16. Broadening angle α as a function of the increase in oxygen flow. Depending on the balance of forces involved three different broadening mechanisms can be distinguished, as discussed in the text.

influence the infrared absorption of the gas and therefore the radiation loss of the crystal.

The formation of the excess melt necessary for broadening requires a certain amount of energy. However, this energy is set free again upon crystallization. Therefore, the end of a broadening step is defined by the power required to compensate for the increased radiation loss, which is caused by the enlarged melt and crystal surfaces.

The theory of the float zone process[35,36] cannot be applied to the Verneuil process.[37] Essential assumptions of that theory are that the liquid is held between two crystal rods, and that the liquid–solid interface is horizontal; both do not apply to Verneuil growth. In addition, interface tension and the most important transport of melt and heat are not formulated at all in zone refining theory.

This outline should be a stimulus and starting point for formulating a full theory of the Verneuil process. Goals of the theory would be, for example, to define the threshold values A and B (Fig. 16) and to provide information on the thickness of the melt film. To support the theory it would be desirable to develop some means for measuring the dynamic behavior of the melt film thickness.

An initial thermal model has been developed by Khambatta *et al.*[8] A thermal balance is set up in which

latent heat + heat transferred to the crystal from the melt
= heat conducted away from the interface

Evaluation leads to a slope showing the crystal retraction rate as a function of the muffle wall temperature. If the growth rate would be controlled only by the rate at which the latent heat is removed by the crystallizing zone, then a temperature difference between muffle wall and solid–liquid interface of only 50°C would suffice for a growth rate of 2.5 cm/hr. The model does not allow one to estimate melt film thickness or the importance of the different growth mechanisms mentioned.

5. THE COOLING PROCESS

As already mentioned the chief advantage of the Verneuil method, the fact that no crucible is required, is of particular significance for the manufacture of materials with very high melting points. A high melting point means, however, that a correspondingly great temperature range has to be passed through before the material has cooled to room temperature; in the case of substances such as sapphire, ruby, or Mg–Al spinel this is about 2000°C. The thermal stresses produced during the cooling process impair not only the crystal quality[38] but above all the mechanical strength of the crystals. Therefore, if high-grade crystals are required, i.e., crystals of low defect density and low stress, the cooling process becomes just as important as the growth process.

5.1. Temperature Conditions during Cooling

The axial temperature gradient in a sapphire crystal has been measured by Ordway and Miller[14] during growth. They used an Ir-60:Ir-40 Rh thermocouple mounted on the seed rods. As growth proceeded the junctions became surrounded by the melt and later by the crystal. Just below the liquid–solid interface a gradient of 50 to 150°C/mm was found, followed by a logarithmic decrease of 100 to 150°C over the first 4 to 5 mm. After this, the gradient remained constant at 6 to 8° per mm. Shiroki[15] measured the axial temperature gradient by inserting a Pt–Rh 20/40 thermocouple into a ruby crystal, putting the crystal back into the furnace, and heating it up for 3 hr. The temperature distribution found with and without additional heaters (see Section 6.1.2) is shown in Fig. 17a. It shows ~350° on the first millimeter, ~50° along the following 4 mm, the lowest decrease being ~10°/mm as found by Ordway *et al.*[14]

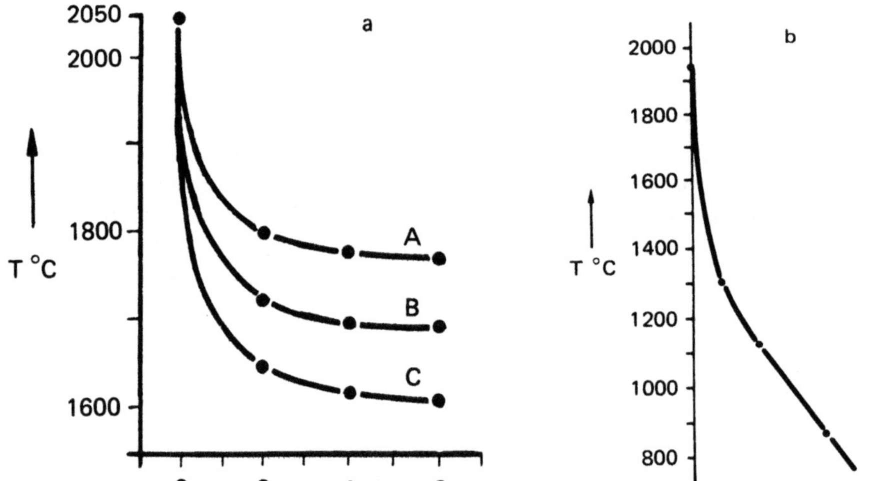

FIGURE 17. (a) Effect of a direct lateral gas heater on the axial temperature gradient in ruby, as measured by Shiroki.[15] The diameter of the crystal is 15 mm. A, Intensive heating; B, average heating; C, no additional heating. (b) Decrease of temperature inside a well-insulated furnace after the flame has been quenched, as measured through the viewing port by a pyrometer.

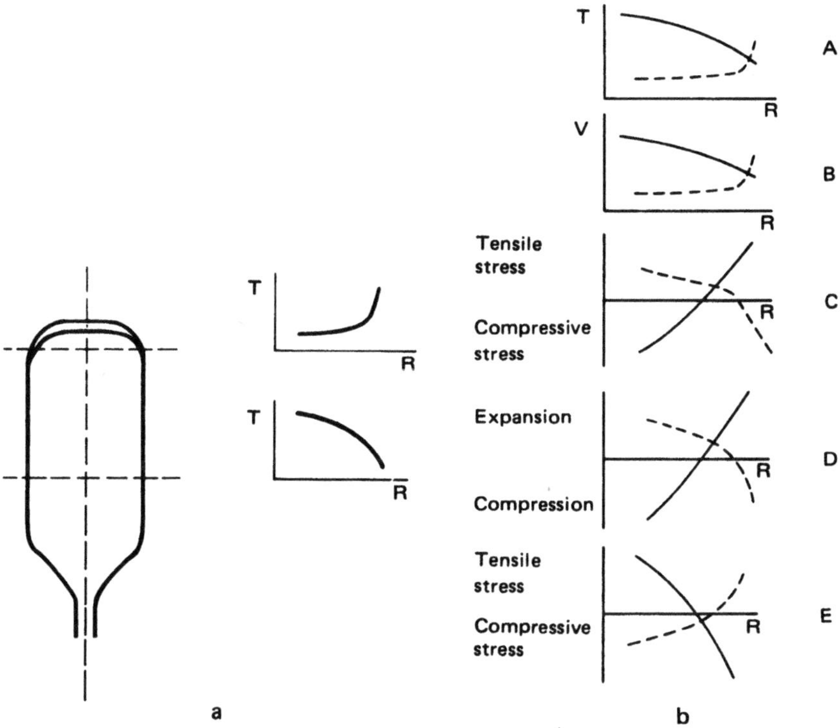

FIGURE 18. (a) Nonuniform cooling in radial direction in the top zone of the crystal (cooling mechanism I) and at a distance from the top or after the flame has been extinguished (cooling mechanism II) as discussed in the text. (b) Development of circumferential stress during cooling. Mechanism I: (---), Mechanism II: (——). (A) Temperature distribution along crystal radius. (B) Temperature dependence of specific volume V ($V = 1/\rho$, where ρ is density). (C) Circumferential stress arising due to $V = V(T)$. (D) Plastic deformation arising due to the stress. (E) Stress in the cooled-down crystal due to the plastic deformation at high temperatures.

Thus it can be stated that there is a very steep axial gradient in the immediate vicinity of the solid–liquid interface decreasing by an order of magnitude within 0.5 cm from the interface. The difference in the measurements may be due to different furnace insulation. In addition there occurs another sharp decrease of the temperature when the flame is extinguished (Fig. 17b). In both cases the strong decrease in temperature may be explained by the factors which contribute to the heat loss. Radiation increases with the fourth power of temperature, heat conduction only linearly.

Measurements of the radial temperature gradient in sapphire have been conducted by Shiroki[15] in the same way as described earlier. They show that during quasi-growth conditions the radial temperature decrease is very small within the core of the crystal. Unfortunately, the temperature distribution in the outer parts of the crystal has not been measured. As photoelastic investigations show, this is the area in which the heaviest stresses form. Figure 18a shows two ways of nonuniform cooling in a radial direction. A radial temperature gradient, in the sense that the temperature increases toward the outside during growth, exists in the outer zone of molten top due to the effect of the flame: The edges of the upper crystal end are rounded off (the liquid–solid boundary line is at least 5 mm lower than the topmost point of a sapphire crystal). As the distance from the top increases and cooling proceeds to lower temperatures the radial temperature gradient in the outer parts will become inverted due to heat loss from the surface by radiation and convection.

5.2. Mechanisms Inducing Stress

The cooling process begins with solidification at the liquid–solid interface during the growth process. As temperature decreases, two contrasting and oppositely acting cooling mechanisms become active (Fig. 18b):

Mechanism I. At high temperatures, in the vicinity of the growth front, it can be argued that the presence of the curved-down part of the melt film implies a horizontal *radial* temperature gradient in contrast to the vertical temperature gradient near the axis. (Note that anisotropy in thermal conductivity in noncubic materials could complicate matters still further.) The solid zone bordering that part of the liquid which is curved downward at the sides must therefore cool more slowly than the core zone with its initially high axial temperature gradient. Thus the core contracts away from the outer zones. The resulting stresses activate slip systems, and plastic deformation occurs. This mechanism operates under growth conditions.

Mechanism II. The growing crystal is progressively lowered and the grown material gradually becomes removed from direct exposure to the flame on its outer surface. As this happens the radial temperature gradient of mechanism I becomes reduced, and eventually inverts, which results in a stress distribution opposite to that characteristic for mechanism I. By radiation and convection, the outer parts of the crystal cool down faster than the core, and the temperature decreases radially outward from the axis. This causes the outer parts to contract more than the core. The plastic deformation which develops because of the resulting stress system leads to compression of the core and tension in the outer zones. Cooling mechanism II is generally known in the handling of metals[39] and can be applied to a number of crystals, as, e.g., zone-refined

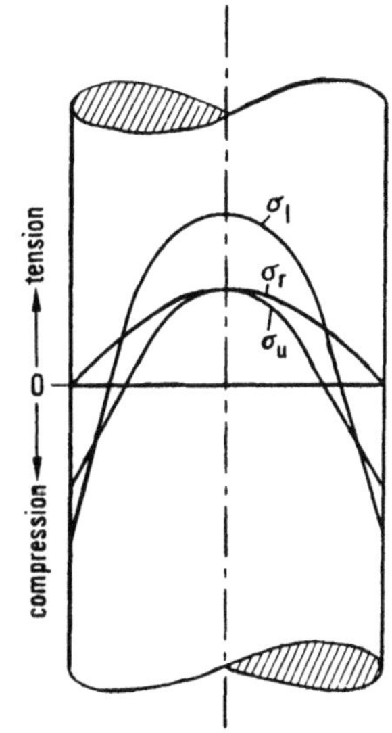

FIGURE 19. Stresses in a cooled-down cylinder: σ_l axial, σ_r radial, σ_u circumferential (cooling mechanism II, see text). (After Stroppe.[39])

silicon[40] and ionic crystals.[41,42] Closer consideration shows that distinction must be made between axial, radial, and circumferential stresses (Fig. 19). Of these, circumferential stresses are considered to be the most dangerous. Calculations have been carried out for ferrites by Shchelkotunov and Danilov.[9]

Stresses, particularly axial stresses, which develop according to mechanism I, have been found by various authors who investigated sapphire or ruby slices cut parallel to the growth axis.[33,43,44] Non-stoichiometric spinel crystals in contrast show circumferential and axial stresses which are caused by Mechanism II.

A third mechanism has been suggested by Shiroki.[15] By considering essentially the high-temperature region, which is also a highly plastic region, he assumes the radial temperature gradient to be small in comparison to the axial gradient. The crystal is thought to be divided into thin horizontal layers; a high temperature is attributed to the upper and a relatively low temperature to the lower of two adjacent layers. Compressive stress then acts on the upper, tensile stress on the lower layer. The resulting stress is described as radial. These radial—and probably also circumferential—stresses should increase from the seed end to the top of a crystal.

No generally applicable rule can be given as to the stresses and strains to be expected with cooled-down Verneuil crystals. Plastic deformation depends on the isotropy or anisotropy of the slip systems as well as on the temperature and on the temperature range in which they are active. [From annealing experiments with corundum (Section 8[119]), it is known that plastic deformation below, say, 1600°C can be neglected over times of the magnitude of a growth period and for stresses as prevailing in cooled-down corundum crystals.] In addition, the orientation of a crystal and its relation to the different slip systems are important. Thus it could well be possible that the temperature and stress inversion during the cooling process leads to a change of the strains as cooling mechanism I ceases and mechanism II comes into action.

5.3. Stress Determination

A method often used in technology for investigating locked-in stresses makes use of the principle of photoelasticity.[45] The usual procedure is to make identically shaped models from a suitable material. These are then subjected to the desired stress and photoelastically examined. Figure 20a shows the stress pattern in a model during cooling, Fig. 20b that in an actual Mg–Al spinel crystal. The model evidently provides a picture similar to the stress situation in the crystal. (At the start of the experiment the temperature of the model material was 120°C. Stress was applied by cooling down to room temperature over a period of 10 min.) The stress cross of the crystal was due to such a weak stress condition that it could only be rendered visible in the little stressed core of a 40-mm-long piece. Owing to the low loadability of the model material while hot, it was not possible to generate high-stress zones. In the following, therefore, direct photoelastic observation of the cooled crystals is described. For such investigations, optically isotropic crystals are naturally advantageous; the following examples thus refer to cubic Mg–Al crystals.

Anisotropic crystals such as ruby and sapphire have also been investigated.[32,43,46,47] For investigation, such crystals need either to be grown in the direction of the optical axis or observed in this direction. Because of superimposition of the natural birefringence and the stress

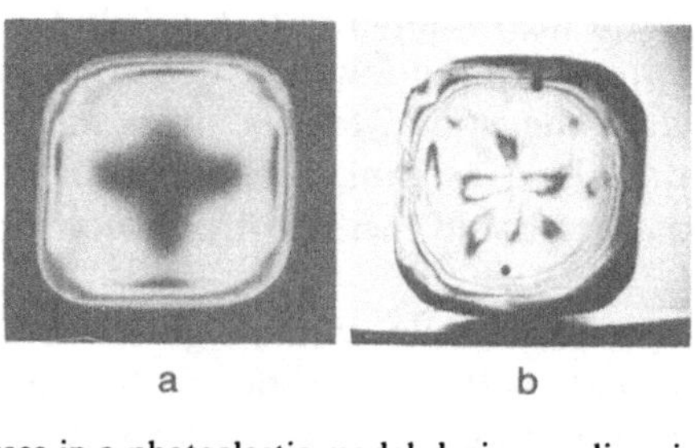

a b

FIGURE 20. (a) Stresses in a photoelastic model during cooling; initial temperature 120°C (see text). (b) Stresses in a Mg–Al spinel crystal; length 40 mm. ×1.

birefringence, some distortion is bound to occur. Difficulty arises when the crystals are not transparent; in such cases, infrared radiation may be tried.[48] Mathematical treatments of stress distribution in cylinders have been given by Sirotin,[49] Parsons,[50] and O'Rourke and Saenz.[51]

Photoelastic Method

Every stress can be represented as the vector sum of two principal stresses, one perpendicular to the other. Polarized light impinging on a stressed crystal is split up into two components whose directions coincide with the two perpendicular directions of the principal stresses T_1 and T_2 (stress birefringence). The phase difference S in wavelengths between the components of the light beam is proportional to the difference between the principal stresses

$$S = \frac{cd}{\lambda}(T_2 - T_1) \tag{1}$$

and can also be represented as the difference between two refractive indices n

$$S = \frac{d}{\lambda}(n_2 - n_1) \tag{2}$$

so that

$$n_2 - n_1 = c(T_2 - T_1) \tag{3}$$

where d is the thickness of the specimen, λ is the wavelength, and c is a constant. The stress birefringence is rendered visible in the photoelastic test device, which in its simplest form consists of a light source, a polarizer, and an analyzer. If monochromatic light is used, dark lines (isochromatics) appear in the presence of stress birefringence (Fig. 21a). Along an isochromatic, the difference between the principal stresses is constant. If white light is used, colored lines appear in the order of the interference colors; the zeroth-order isochromatic appears dark (Fig. 21b). With increasing orders, the stress difference [Eq. (1)] increases. Dark lines, the isoclinal lines, are superimposed on the isochromatics. On an isoclinal line, the principal stresses run parallel to the directions of polarization. The isoclinal lines can be eliminated by inserting quarter-wave plates behind the polarizer and in front of the analyzer. The principal directions of the first plate are rotated through 45° against the direction of polarization, those of the second plate through 90° from those of the first plate.

The number of isochromatics observed may depend on the thickness of the specimen. Weak locked-in stresses only become visible in the isochromatic pattern when the specimen is sufficiently long. In such cases, by using short specimens, it is possible to practically suppress the isochromatic pattern, thus obtaining the pure isoclinal pattern.

Figure 21 shows the isochromatic pattern of a specimen cut from a Mg–Al spinel perpendicularly to the growth direction ⟨100⟩; its length is 15 mm. The isochromatics run parallel to the lateral faces of the crystal in the peripheral zone. If the piece of crystal is split along a plane through the growth axis, the isochromatics essentially disappear except for those in the corners of the crystal; i.e., the stresses are of an elastic nature.(31) The plastic deformation which initially gave rise to the stress will of course persist.

Figure 22 shows the isochromatic pattern in the longitudinal section of a Mg–Al spinel crystal. Periodic axial stress is indicated by the zero-order isochromatics running parallel to the growth front. Axial stresses at the crystal edge are concentrated in a thin layer. As piezooptical investigation shows (see later) compressive axial stress in the edge zone is counterbalanced by tensile stress in the core (Fig. 25).

The isoclinal lines provide information about the direction of the stresses. Figure 23 shows examples of isoclinal lines such as may be observed in Mg–Al spinel wafers between crossed polarizers. The specimens were cut from the crystals perpendicularly to the growth direction. They are 1 mm thick and originate from different parts of crystals and from crystals with different composition. The photographs in series (b) are of the same specimens as series (a) but rotated through 45°. The stresses are seen to be radial and circumferential. The differences between series (a) and (b) are due to the fourfold symmetry of crystal shape and to (111) slip planes; it should not occur in rods of perfect cylindrical symmetry.

→

FIGURE 21. Isochromatics in a Mg–Al spinel specimen cut from the crystal perpendicularly to the growth direction ⟨100⟩. Length of specimen 15 mm; composition $MgO:Al_2O_3 = 1:2$. (a) Monochromatic light, (b) white light. Both ×1.

FIGURE 22. Isochromatics in a longitudinal section of a Mg–Al spinel ×1.

FIGURE 24. Single crystal bar of Mg–Al spinel bent by action of screw (indicated by arrow) and viewed between crossed polarizers, with gypsum plate on top of sample. Zone of compressive stress blue, of tension yellow. Supporting knife-edge bearings at the extreme ends of the lower side are not shown.

FIGURE 25. (a) Compressive circumferential stress at the edge zone of a slice cut perpendicular to the growth axis; this disappears after fracture. (b) Compressive axial stress at the edge zone and tensile axial stress in the inner parts of a slice cut parallel to the growth axis. Shown is only a small, but typical, section (×3.5); direction of growth is upwards and parallel to the shorter dimension of the figure.

FIGURE 21a

FIGURE 21b

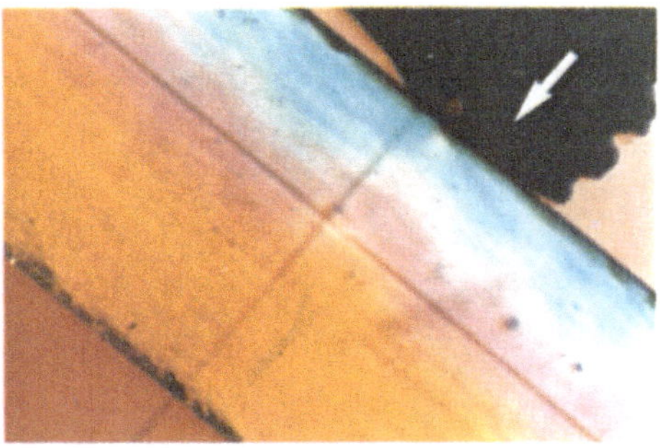

FIGURE 24

FIGURE 25a

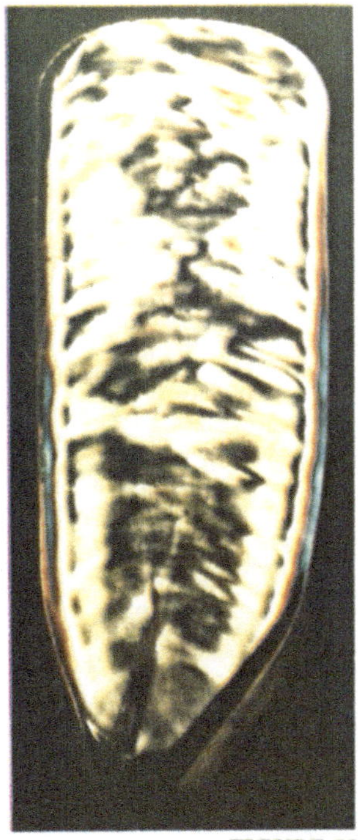

FIGURE 22

FIGURE 25b

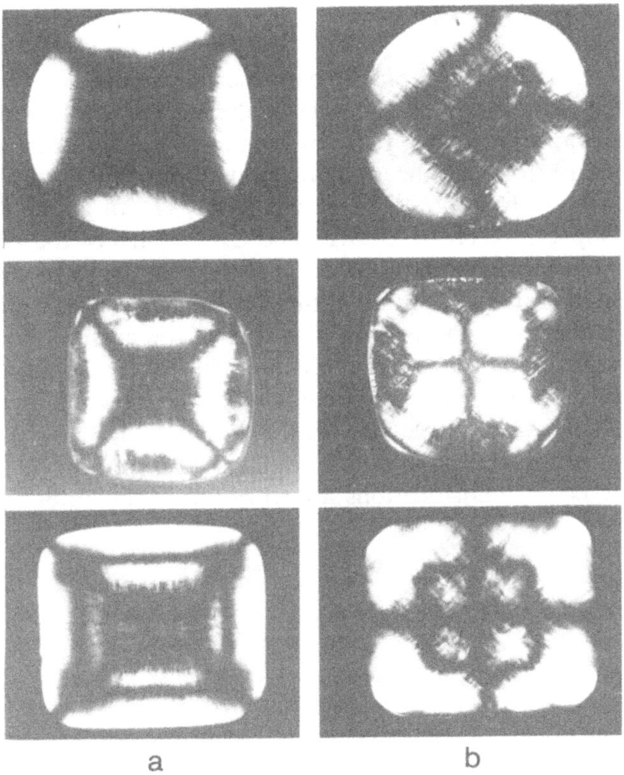

FIGURE 23. Isoclinal lines in Mg–Al spinel wafers. Plane perpendicular to the growth axis ⟨100⟩. Specimens 1 mm thick, diameter ~20 mm. In series b, the wafers have been rotated through 45° as compared to series a.

It is important to ascertain where the compressive and the tensile stresses are to be found. To determine this, a compensator, e.g., a gypsum plate ("red, first order"), is placed on top of the specimen, which is sandwiched between crossed polarizers. From the color shifts obtained through subtraction or addition of the phase difference, stresses of different sign can be deduced. Calibration is possible by observing the color shift when a bar-shaped specimen is bent. Figures 24 and 25 show examples of Mg–Al spinel specimens. From Fig. 24 compression is seen to be characterized by blue, tension by yellow color (the color distribution depends on the angle of rotation of the sample relative to the polarization direction and is not generally valid). Figure 25a shows compression at the edge zone of a slice cut perpendicular to the growth axis. The distribution of tensile and compressive axial stresses in a slice cut parallel to the growth axis is shown in Fig. 25b.

5.4. Plastic Deformation

As already described plastic deformation occurs due to unfavorable temperature conditions in the crystal during the cooling process. It is marked by the occurrence of slip or twinning or, in a more irregular way, by the formation or sliding of grain boundaries.[52] In sapphire two slip systems can be observed:[44] basal slip {0001} ⟨11$\bar{2}$0⟩ above 900°C,[53] and prismatic slip {1$\bar{2}$10} ⟨$\bar{1}$010⟩ above 1800°C.[54] Which of the two systems is activated, and in which phase of the growth or cooling process this occurs, depends on the relative position of the slip direction and the direction of the stress difference: i.e., the growth direction of the crystal also plays a part. Davies investigated dislocation arrays by etching.[44] He found the activation of both systems in a sapphire crystal, for which the growth direction was at an angle of about 70° to the *c*-axis (Fig. 26).

In Mg–Al spinel crystals the slip systems {110}⟨110⟩ and {111}⟨110⟩ have been observed.[55–57] Dislocations are frozen-in in the outer regions. In the

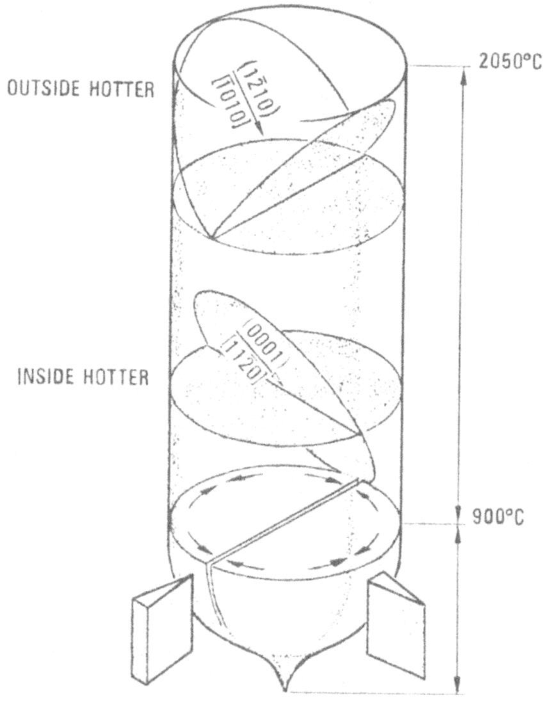

FIGURE 26. Schematic representation of the conditions leading to basal and prismatic slip in sapphire and which may result in cracking of the crystal in the way shown, because of anisotropic stresses. (After Davies.[44])

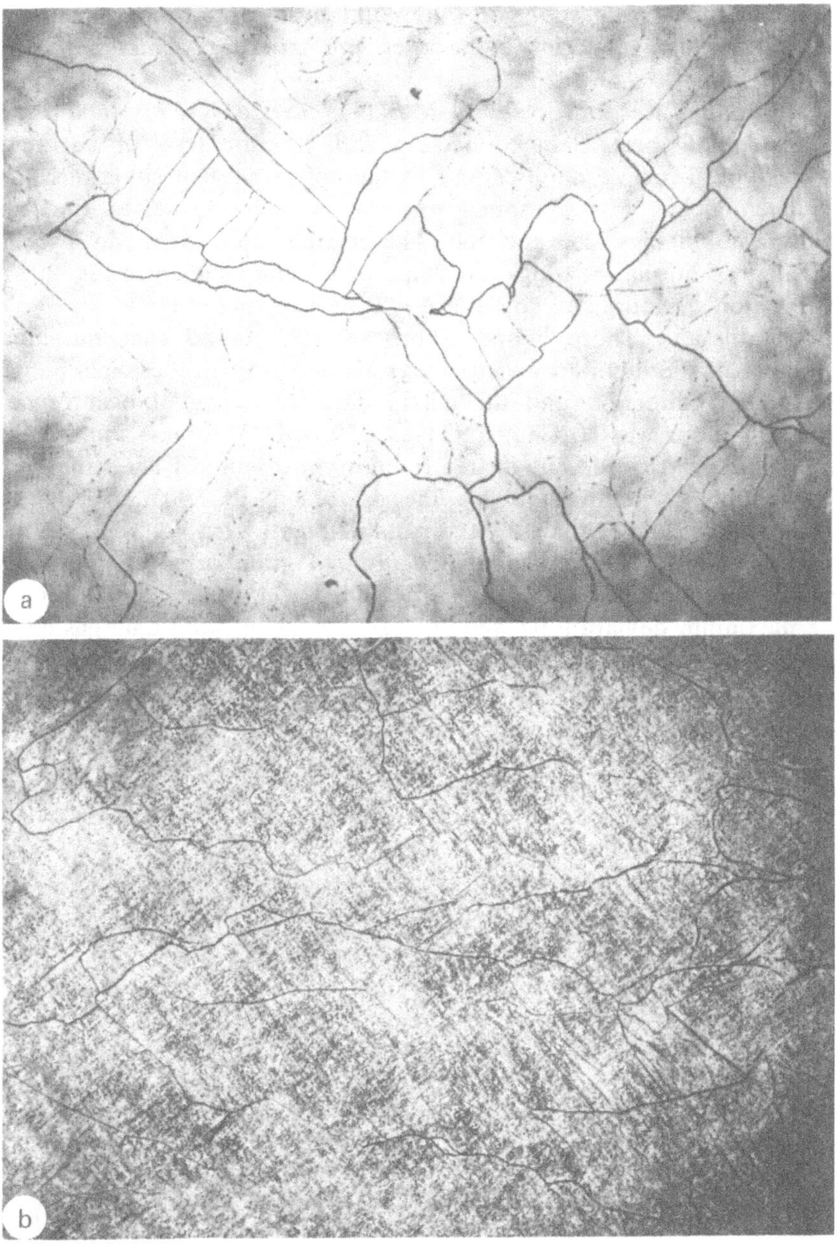

FIGURE 27. Etched {100} plane of Mg–Al spinel wafer cut perpendicular to the ⟨100⟩ growth direction, showing slip parallel to the {111} plane. (a) Core area, (b) area close to the edge of the crystal.

interior of the crystal, polygonization and grain formation are found (see Fig. 27). The etch pit density in the outer regions of the crystals is of the order of $10^5\ cm^{-2}$.

Twinning was observed by Merker[58] in $CaTiO_3$ crystals, and by Heuer and Roberts[59] in connection with fracture tests on corundum. Quite typical for Verneuil crystals is the appearance of subgrains. Their arrangement in Mg–Al spinel and sapphire is shown in the X-ray topographs in Figs. 28a and 28b. The crystals were grown in a furnace (without additional heating) the shape of which is shown in Fig. 30. The orientation deviations in spinel are $<2°$, in sapphire $<1°$.[60]

Curtis and Thorp[61] and Thorp *et al.*[62] found that the mosaic structure of Verneuil-grown ruby crystals depends on the angle between the growth direction and the *c*-axis. The frequency of occurrence of mosaic regions was 70% for 0° rubies, 33% for 73° rubies, and only 11% for 90° rubies. (Percentage is area fraction misoriented.) The value of this information is diminished by the fact that the 90° crystals were considerably smaller than the 0° crystals (length 8 cm vs. 3 cm, diameter 2 cm vs. 1 cm). The maximum deviation of grain orientation from the growth direction was found to be 2°. Shiroki[15,63] succeeded in reducing the maximum deviation between mosaic blocks by additional gas heating of the side walls of the crystal (see also Section 6.1.2.2).

5.5. Fracture

As has been shown in the preceding sections, stresses arising during the cooling process lead to plastic deformation which in turn gives rise to

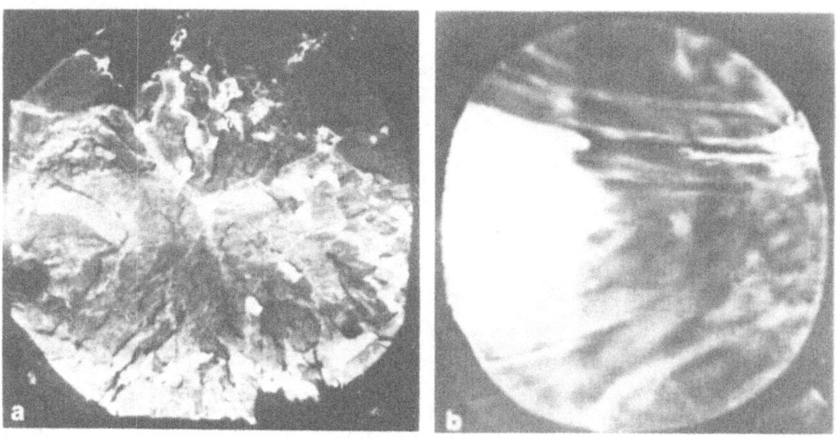

FIGURE 28. X-ray topographs (Lang method) of (a) Mg–Al spinel and (b) sapphire, the reflecting planes being {400} and {012}, respectively. Radiation Mo $K\alpha_1$. (From Falckenberg,[60] reprinted by permission of the Electrochemical Society, Inc.)

stress in the cooled-down crystal. If no special precautions are taken, especially in respect to the insulation of the growth furnace, mechanical strength of the crystals may be low. Thus, a high proportion of the crystals produced in factories on a large scale are delivered in the form of split boule halves; usually the boules are halved by just tapping the crystal top, assuming that the crystals did not split during cooling. Like plastic deformation, the yield of unfractured crystals depends also on crystal orientation.[64]

The way corundum crystals crack was investigated by Indenbom and Tomilovskii[33] and by Davies.[44] Boules and, particularly, rods split along the growth axis. If the halves were put together and compressed on one end, a gap would open toward the opposite end. From the magnitude of this gap a stress of 200 to 900 kg/cm^2 was calculated.[33] From the way the crystal halves were bent it was concluded that axial tension prevails in the outer zones, axial compression in the internal part of the unfractured crystals. From Fig. 26 it can be seen that stresses of this kind are due to plastic deformation which develops at high temperatures in the top zone of a crystal.

Sometimes it can be observed that corundum or Mg–Al spinel crystals crack without apparent cause perhaps days or even weeks after they have been grown. Since investigations by Wiederhorn[65] have shown that plastic deformation by dislocation motion or twin formation plays no role at temperatures below 400°C,[66] one of the following assumptions may offer an explanation:

1. Residual stress may induce diffusion of vacancies toward the crack tip.
2. Fracture may result from atomic bond fluctuations at the crack tip.
3. Fracture may be controlled by rearrangement of bonds or free radicals formed during the fracture process.

Fracture of corundum single crystals usually takes place along certain fracture planes, for example, $\{10\bar{1}1\}$, $\{11\bar{2}0\}$, and $\{10\bar{1}0\}$. Bagdasarov *et al.*[67] were able to show that $\{10\bar{1}1\}$ and $\{11\bar{2}0\}$ tended to be the more probable cleavage planes.

5.6. Remarks on the Cooling Process

The reasons for fracture as well as for plastic deformation are the quenched-in stresses which are due to unfavorable temperature conditions during cooling. Anything that lowers radial and axial temperature gradients in the crystal will also lower these stresses. Two general approaches are to improve the method of cooling and to modify the design of burner and furnace.

There are several methods for cooling crystals: the flame may be extinguished slowly or suddenly, the crystal may be withdrawn slowly from the flame, or the furnace may be opened. The last method is standard in the final cooling phase of the fabrication of crystals of small diameter in order to speed up the output. The withdrawal method can be important in the manufacture of crystals for which precipitation strengthening occurs on cooling (Section 8). This is the case, for example, with Mg–Al spinel with a high Al_2O_3 content. The method in which the oxygen and hydrogen supplies are cut off suddenly and simultaneously has in some cases proved to be superior to the method in which the flame is extinguished slowly. This at first sight unexpected behavior is due to the fact that slowly extinguishing the flame is connected with a reduction in the length of the flame, which leads to nonuniform cooling of the crystal.

All these methods of cooling are unsatisfactory if crystals with a particularly high melting point or a large diameter (e.g., for sapphire $\geqslant$25 mm) are to be manufactured. As pointed out in Section 2, there is a close interaction between burner, furnace, and crystal. In particular the interchange of radiant heat between the crystal and the furnace wall influences radial as well as axial temperature gradients in the crystal. Additional heating and alteration of the flame guidance have therefore been introduced, and are now discussed.

6. GROWTH PARAMETERS

6.1. Growth Chambers

Growth chambers of various designs, each used to grow a certain crystal or group of crystals, have been described in the literature. In the interests of further development of the process, it is obvious that an assessment of these experimental designs is needed which is as far as possible generally valid and which is not restricted to a particular design or result. Basically the designs are attempts to solve the same problem, namely, to have control over the temperature gradients in the crystal-growing chamber. Quality and size of the crystals grown are certainly criteria for evaluating a particular apparatus, although the dissimilarity of the crystals and differences in desired crystal size have to be considered. A further criterion is the general practicality of the particular furnace design.

It must be stressed that temperatures of 2000°C and above in conjunction with the flame atmosphere pose problems which, because of the materials available, are at the borderline for a simple technical solution. The insufficient thermal refractoriness of those materials that are available sets marked limitations on furnace design and construction (see Section 6.1.3).

6.1.1. Simple Furnaces

Tubular furnaces provide the simplest type of growth chamber. In constructing them special attention must be paid to insulation quality and to stability at temperature.

Industrial furnaces are usually of simple and relatively cheap construction. Figure 29a shows a typical crystal-growing furnace used for crystal manufacture on a large scale. It consists of two interlocking halves; a tongue and groove device seals off the growth chamber from its environment. The crystal-growing process can be observed through a viewing port. The furnace material is usually sintered alumina. The mean lifetime of such a furnace is 10 to 20 crystal-growing processes. The crystals grown in this way, for example rubies, have a typical diameter of up to 15 mm and a length of around 50 mm. Because of the relatively poor thermal insulation temperature gradients tend to be high and, as a result, the mechanical stability of the crystals grown is frequently poor.

The laboratory furnace shown in Fig. 29b provides much better results than industrial furnaces. The main structural feature is its improved insulation, which is achieved by employing better quality materials in greater thicknesses. The growth chamber consists of densely

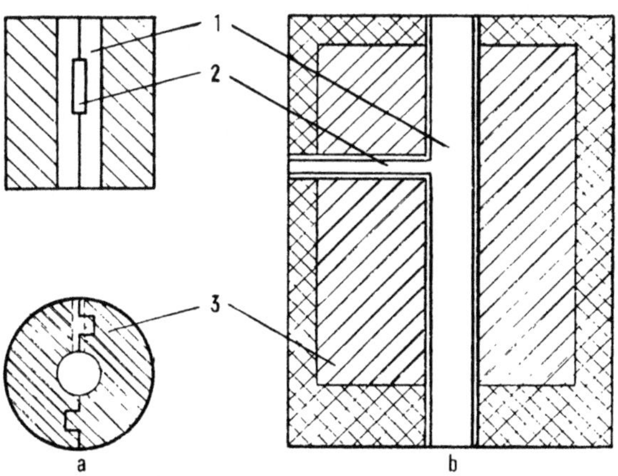

FIGURE 29. (a) Typical industrial furnace (made in two halves clamped together) for growing corundum used as gem stones and bearings. (b) Laboratory furnace with highly insulated walls. 1, Growth chamber; 2, viewing port; 3, heat insulation. Scale ~1:8.

sintered 99.7% alumina tube. This is in turn enclosed in materials of different thermal refractoriness and of minimum heat conductivity in each case.

The process is observed through a 99.7% alumina tube with an ~10-mm bore. The temperature at the inside wall at the level of the viewing port is about 1800°C compared with around 1400°C in a simple industrial furnace of the type just described. In a furnace of this kind rubies with a length of 250 mm and a diameter of around 25 mm have been grown. An improved version in which the process is viewed from above, through the burner, has been described by Seifert.[68] With this, it is possible to avoid asymmetries of the temperature distribution in the furnace which would be due to the usual lateral arrangement of the viewing port. For the same purpose, Shiroki[69] closed the viewing port after having finished the broadening process; further growth was controlled by a thermocouple measuring the temperature in the furnace at a point close to the growing crystal. However, it should be mentioned that there is no investigation of asymmetric stresses in crystals which has shown them to be caused by the viewing port.

Using the ordinary laboratory furnace mechanically stable sapphires with a diameter of about 30 mm can be manufactured. The maximum diameter that can be achieved, but with crystals of poor strength (e.g., showing fracture during cooling), is considerably higher. Sapphires with a diameter of up to 80 mm were obtained in an apparatus with an additional adaptor tube (Fig. 35). This is a 99.7% alumina tube which reduces the effective diameter of the growth chamber. It extends up to the level of the viewing port, acts as a radiation reflector, and is withdrawn downward continuously from the furnace as the crystal diameter approaches the inner diameter of the adaptor tube. The inner diameter of the adaptor tube was approximately 70 mm in this case, the diameter of the growth chamber 90 mm (see Table 3). It should be noted that it is naturally only possible to obtain such large crystal diameters if a multitube burner is employed, as is discussed in Section 6.2, since only with this can a wide enough flame be obtained to avoid melt runover (see Fig. 34d).

6.1.2. Growth Chambers with Additional Heating

There are three points to consider in connection with additional heaters. These are the maximum temperature the whole device can withstand, the way the additional heater is arranged in the growth furnace, and the power required. Experimentally it has been found that in well-insulated furnaces the power may be less than half that of the main burner. The geometrical arrangement of the heater is obtained from the considerations in Sections 4 and 5, i.e., it should give a small radial gradient at the height of the growth front and a small axial gradient below that position in the furnace.

The first requirement is very difficult to meet because additional heating of the side walls of the growth chamber at this height will lower the liquid–solid boundary line. This means that the zone where cooling mechanism I is active (see Section 5.2) will be extended instead of shortened. Taking this into account one of the following two approaches must be adopted:

1. The subsidiary heating zone is extended far below the growth front. Consequently, the axial gradient near the growth front becomes low; the temperature difference between core and rim must then become lower, too. In addition a certain amount of annealing occurs in the lower parts of the crystal which will decrease the stresses due to plastic deformation.
2. Means are found for flattening the growth front, i.e., how to lift up the liquid–solid boundary line relative to the crystal top without simultaneously cooling the bulk of the crystal.

Furnaces in which one or the other of these has been realized or at least approached have been designed, and are discussed in the following.

The most important and, at the same time, the most critical factor is the maximum temperature the overall apparatus can withstand. It is essential that the growth chamber withstand the double load of burner flame and additional heater at the desired maximum temperature. Two methods for providing subsidiary heating are available: electricity or gas.

6.1.2.1. Subsidiary Electrical Heating. With an electric heater it is necessary that the heating elements are chemically and thermally stable at operating temperature. An additional electric heater was used for the first time by Merker in 1962,[58] who grew $CaTiO_3$ crystals (melting point around 1950°C). His reason was that the crystals cracked during the rapid cooling process inevitably with a simple tubular furnace (the crystals cooled down from 1400 to 500°C in 5 min). Merker used a heating coil made of platinum wire, which was built into the wall of the lower part of the furnace. To reduce the axial gradient the afterheater was kept at 1400°C during growth. When the growth process was completed, the crystal was withdrawn into the additional heated zone, the heater was set to the maximum temperature obtainable, 1550°C, and the flame was extinguished. The crystal temperature was then maintained at around 1480°C for 48 hr; subsequent cooling to room temperature took place at a rate of 30°/hr. Apparently because of the rapid initial cooling from 1950 to 1550° there was so much residual stress in the crystals that they cracked on machining. It was possible to eliminate this stress by annealing the crystals in a gas-heated furnace at 1700°C for 6 hr. The rate of cooling was then 50°/hr.

Heating elements made of Superkanthal were tested by Mitchell,[70] but failed after being used for 10 to 15 hr at 1700°C due to the reducing effect of the furnace atmosphere.

Arlett and Robbins[71] used a Pt–10% Rh coil for the growth and *in situ* annealing of stoichiometric spinel, $MgAl_2O_4$ (melting point 2150°C). In this case too the coil was in the lower part of the furnace and was kept at 1450°C during the growth process. The crystal was then withdrawn into the additionally heated zone and cooled from 1450 to 400°C at a rate of 10°/hr. The crystals did not crack during cooling. The furnace temperature values measured indicate that the wall temperature was only 1400°C at the lower edge of the viewing port during the growth process.

Heating elements of Pt–Rh have also been applied by Shankland.[30] In the furnace of Belyaev *et al.*[72] the heating elements were not put close to the growth chamber but were installed in the outer part of the furnace in order to avoid excessive temperatures which would harm the electrical heating elements (see also later).

A tubular furnace with Mo wire coiled over the whole length and sheathed in hydrogen to prevent oxidation was tried by Seifert[68] for the growth of laser rubies. As he pointed out, the lifetime of the furnace was very limited. The present author's later experiments showed that the tube forming the growth chamber cracked; the flame gases which then came through the cracks destroyed the Mo coil. It seems to be more promising to use several rod-shaped heating elements, which can be put into a furnace, each of them being well insulated and thus protected against thermal shock. The rods consist of an alumina tube in which there is a Mo heating wire, which is sheathed in hydrogen during operation.[73] Soviet workers at the Institute of Crystallography of the Academy of Sciences in Moscow have successfully used thick tungsten bar heaters wound around a high-quality impermeable alumina furnace tube, and running in hydrogen.[74] This is discussed further by Goodman in Chapter 4 of the present volume.

One can summarize most experiments with additional electric heaters briefly: If the heating elements are kept at a low temperature, they either fulfill their purpose inadequately or not at all and the crystals crack; if the heating elements are used at the required high temperature, their lifetime is short. Further development would have to concentrate on the improvement of the thermal stability of the heating elements. Perhaps the Soviet work will provide an answer. Alternatively, it may be possible to fabricate a subsidiary heater from stabilized zirconia ceramic: at high temperatures this becomes a good (ionic) conductor.

6.1.2.2. Subsidiary Heating by Gas. An additional gas heater is in many ways better suited to solving the problem since it automatically satisfies the requirement for chemical and thermal stability of the heat source. The temperature can, if required, be set to as high a value as that of the main flame. Such a heater can be used in two different ways: it can heat the crystal either directly or indirectly.

An additional indirect gas heater using an oxygen–propane gas flame was first used by Mitchell[70] for the growth of stoichiometric Mg–Al spinel, ruby, and yttrium–aluminum garnet. Before the growing process started, the furnace was heated up to 200° below the melting point of the material with the aid of the additional heater. The temperature could not be raised further for otherwise the minimum value of the axial temperature gradient required for crystallization during operation was not obtained. Also, under these conditions growth was described as critical in the face of small temperature increases. Upon completion of the growth process the crystal was withdrawn completely into the part of the growth chamber heated by the additional heater, where it was annealed and slowly cooled to room temperature. Further annealing in a suitable high-temperature furnace in air eliminated residual stresses. The results obtained with this arrangement at first appear disappointing, for the crystals had to be postannealed. However, it might be assumed that factors such as the unfavorable geometry of the additional heating zone, which necessitated the withdrawal of the crystal, or the part of the furnace above the crystal, which was made of a material of high thermal conductivity and perhaps brought about an unfavorable temperature distribution, led to the described result.

Pastor and Pastor[75,76] also worked with an additional heater to grow rare earth oxides (melting points between 2250 and 2450°C). The heater operated on the counterflow principle. On account of the high temperatures the furnace interior was made of CaO-stabilized ZrO_2 instead of the usual alumina parts. The resistance to sudden changes of temperature of this material is far poorer than that for alumina, with the result that the parts had to be replaced after every crystal-growing run.

In the following an apparatus is described (Fig. 30) which has been used for the growth of sapphire and Mg–Al spinel crystals of large diameter (⩾50 mm).[7] The flame gases are, as before, applied via a ceramic tube to the growth front of the crystal and transfer the thermal energy to the crystal which is required to maintain the melt film on the top of the crystal. Immediately below the top the flame gases are laterally deflected. This is realized with the aid of a ceramic furnace insert (e.g.,

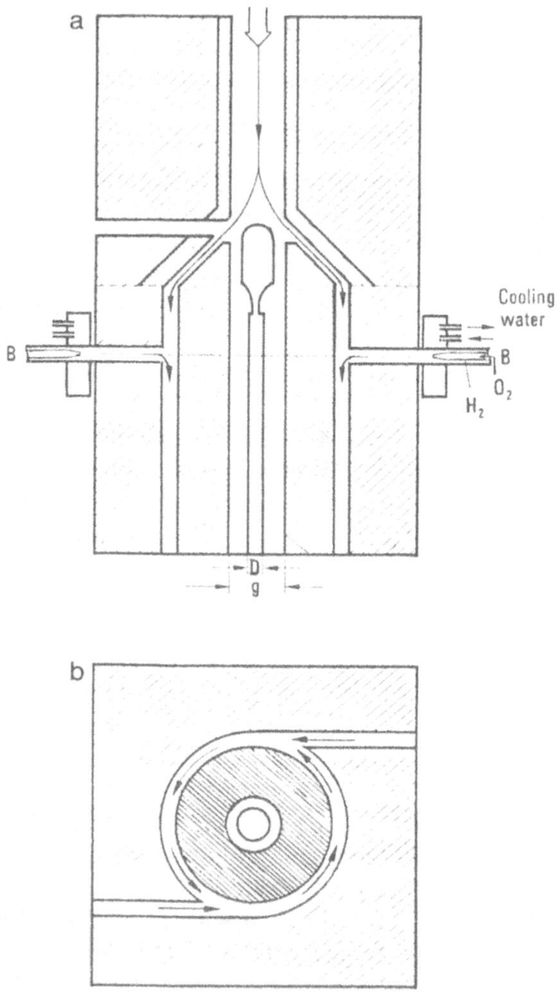

FIGURE 30. · Furnace with ceramic insert and additional gas heating for the growth of mechanically stable sapphire and spinel crystals up to large diameters (~50 mm). Scale ~1:8. (a) Front view, (b) cross section of B–B. For crystals grown in this type of furnace, see Figs. 31 and 34.

alumina), which is shaped like a crucible and encloses the growing crystal without touching it. The diverted flame gases heat the insert. With the exception of the growth front the whole crystal is in thermal equilibrium with the body of the insert right from the start of the crystal-growing process. The thermal shock is therefore small when the flame is extinguished, the cooling rate being largely determined by the thermal characteristics of the insert. When "stress sensitive" crystals are being grown, the insert is heated by a lateral additional heater right from the beginning; cooling takes place by controlled reduction of the additional

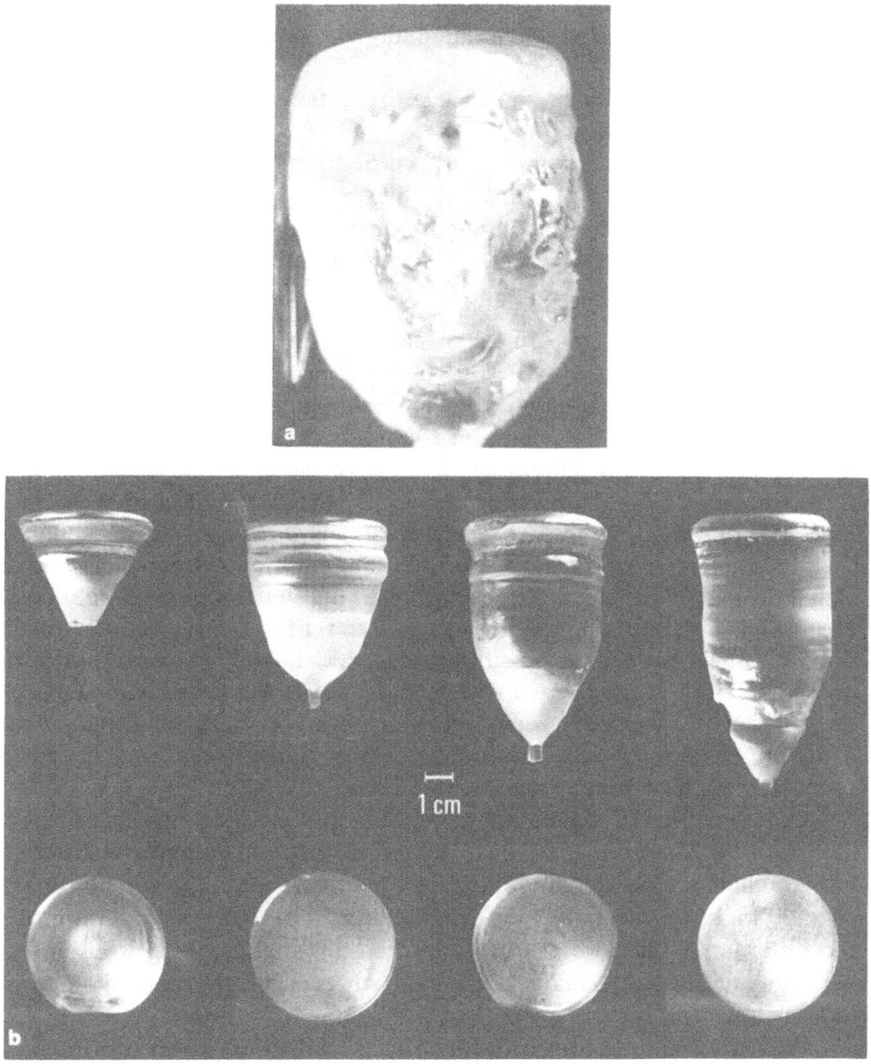

FIGURE 31. Sapphire crystals grown in the furnace shown in Fig. 30. (a) Orientation $\langle \bar{1}010 \rangle$; direct heating of the side faces of the crystal (after Shiroki[15]) has been used in addition to the lateral heaters. Crystal diameter, 56 mm. (b) Orientation $\langle \bar{1}010 \rangle$ (first and third crystal); orientation $\langle \bar{1}012 \rangle$ (second and fourth crystal). No direct heating of the side faces of the crystals. Lower series shows top view of crystals.

heating. As in the case of simple tubular furnaces, an adapter can be used for the growth of crystals of large diameter.[60]

This furnace design has a number of advantages:

1. The liquid–solid boundary line is lifted up relatively to the top of the growing crystal. The effect of cooling mechanism I (Section 5.2) is thus diminished.
2. The thermal shock at the moment the flame is extinguished can be almost entirely avoided.
3. The flame gases of the main burner are simultaneously used to provide subsidiary heating.
4. There is no difficulty in augmenting the additional heating effect of the main burner by separately controllable burners which are placed at the furnace sides.

With this device Mg–Al spinels of composition $MgO:Al_2O_3 = 1:2$ have been grown (see Fig. 34) with a diameter[60] of up to 50 mm and sapphire crystals (Fig. 31) with a diameter of up to 65 mm. The spinel crystals were mechanically stable; with sapphire crystals, however, additional lateral heating was necessary. Direct gas heating of the crystal as described later is also possible (Fig. 31a). An injection of oxygen is sufficient if the main burner is operated, as is customary, with a large excess of hydrogen.

To this type of growth chamber belongs a furnace described by Belyaev *et al.*[72] in which the growth chamber widens immediately below the growth front for a length about equal to that of the crystal. In order to decrease the heat loss a subsidiary electrical heater was installed in the outer part of the furnace. Annealing was done in the lower narrow part of the furnace. Zemlicka[23] reported on sapphire crystals of about 40-mm diameter obtained when cooling was done slowly (regarding the cross-sectional shape of these crystals, see Section 3.2.1).

Another type of additional gas heater was used recently by Shiroki[15] and then by Kvapil *et al.*[16] In this case, instead of the outer walls of the growth chamber being heated, the heating flame impinges directly on the crystal. In both cases, oxygen–hydrogen burners are used. Kvapil used three burners, which were located in the growth plane; Shiroki used four, which were directed against the side surfaces of the crystal near the molten top. The burner axes were horizontal.

The experiments conducted by Shiroki led to the conclusion that the axial temperature gradient can be reduced by the additional heater (Fig. 17); this arrangement led to a reduction of the normally observed misorientation of Verneuil corundum. To increase the mechanical strength, the crystals were subsequently annealed in another furnace.

6.1.3. Refractory Materials for Furnace Construction

Refractory materials are important because the wall temperature of the growth chamber depends to a great extent on the thermal conductivity and thermal capacity of the materials used to construct the furnace. However, their values are never specified in publications on Verneuil growth. In addition the wall temperature and the change in temperature during the cooling period have rarely been measured. It is therefore practically impossible to make any form of comparison of the results in different papers with regard to the same crystal-growing problem.

In assessing furnace materials with regard to their suitability one is very soon faced with the unpleasant fact that the better the high-temperature performance of a material (i.e., the combination of refractoriness and thermal shock resistance), the higher is its thermal conductivity. A well-insulated furnace therefore always consists of at least one highly refractory inner layer and an outer layer of low thermal conductivity. Of the materials in common use Al_2O_3 bricks have the best high-temperature performance. ThO_2, ZrO_2, and MgO can withstand higher temperatures, but have the disadvantage of low mechanical stability; i.e., such bricks break very easily with changes of temperature. Bubble alumina(68) is especially useful; this consists of small hollow spheres which combine high refractoriness with a relatively low thermal conductivity. A number of felt materials have also been developed recently which are stable up to 1500°C with a relatively low thermal conductivity.

Thermal data can be found in a number of books,(77,78) and the compatibility of refractories is discussed, for example, by Ryshkewitch.(79) Finally it can be important to know the temperature distribution within a furnace, and it may be useful to know that Barta *et al.*(80) used the coloring of Fe, Cr, Mo, and W oxides at certain temperatures for making the isotherms visible. (These oxides were incorporated in a mixture of alumina powder and a little phosphoric acid, which was used to prepare the actual refractory parts of the furnace.)

6.2. Flame and Burner

The flame acting on the crystal is produced by the combustion of oxygen and hydrogen. A considerable amount of heat is developed during the reaction:

$$\tfrac{1}{2}O_2 + H_2 \rightarrow H_2O + 57.8\ \text{kcal for steam} \qquad (68.3\ \text{kcal for water})$$

Table 1. Maximum Flame Temperature of Gases Mixed with Pure Oxygen[a]

Gas	Temperature (°C)
Town gas	2000
H_2	2700
CH_4	2700
CO	2925[82]
Acetylene	3100

[a]From Jones.[81]

A number of other gas combinations also release large amounts of energy upon combustion and enable high temperatures to be attained (Table 1). However, it seems to be difficult to obtain a clean combustion with carbon-containing gases even if they are well mixed with oxygen before burning.[83] In the case of the H_2/O_2 flame, the gases are usually mixed outside the nozzles to prevent flame flashback. Results of investigations with acetylene[84] and town gas[85] are reported in Section 9; the redox nature of the H_2/O_2 flame is dealt with in Section 7. Details on laminar diffusion flames may be found in the literature.[34,86–90]

The isotherms in the flame have been measured by Khambatta *et al.*,[8] Bednorz and Scheel,[175] and Ikornikova and Popova.[91] The latter authors used needles of platinum and corundum and plotted the melting point isotherms for these substances. By changing the gas working conditions, details of which were not given, the axial temperature gradient in the flame was flattened out.

Unfortunately, the shape and flow of the flame gases[34] have not been investigated during growth. Such investigations would be of great interest since the shape of the crystal top is defined to a great extent by the flow of and heat transfer from the flame gases. The shape of the molten top in turn influences the strength and plastic deformation of the crystals (see cooling mechanism I, Section 5). Both the directivity and profile of the flame are determined by the design of the burner as is the variation of the flame shape achievable through alteration of the gas flow. A useful rule of thumb is that the effective cross section of the burner should be about the same as that of the desired crystal.[7] Attempts to grow the crystal beyond this diameter may cause melt runover (Section 4.2).

In most cases the burners consist of concentric tubes of stainless steel (Figs. 32 and 33).[92] Oxygen is supplied through the inner tube, hydrogen

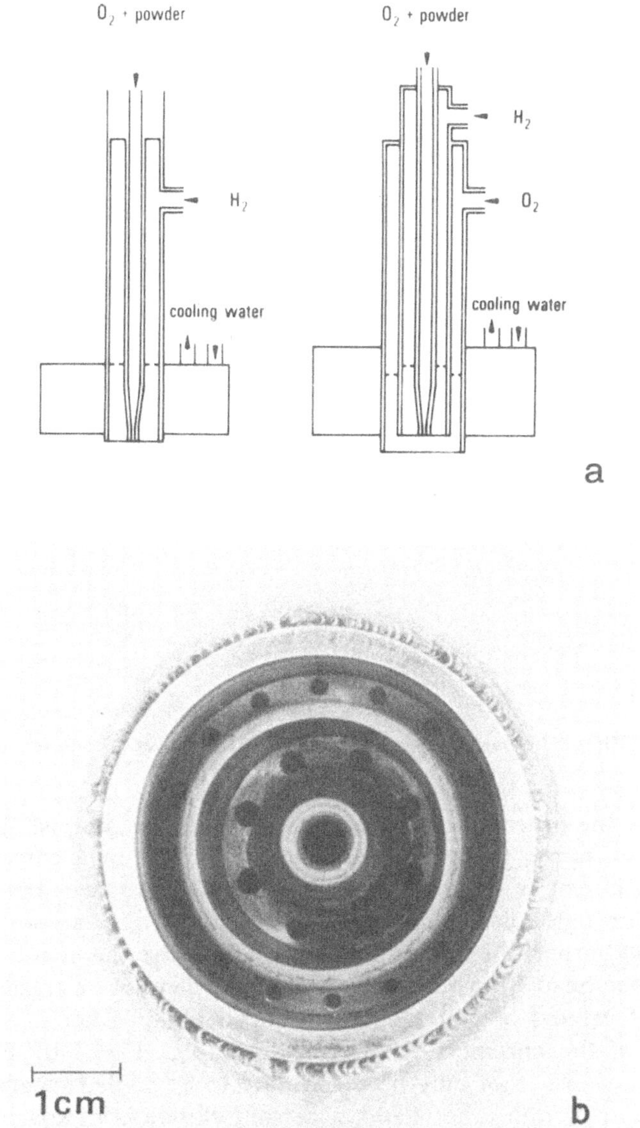

FIGURE 32. (a) Diagram of construction of two-tube and three-tube burners. The dimensions of the two-tube burner correspond to those of the inner part of the three-tube burner.[7] (b) End view of the three-tube burner.

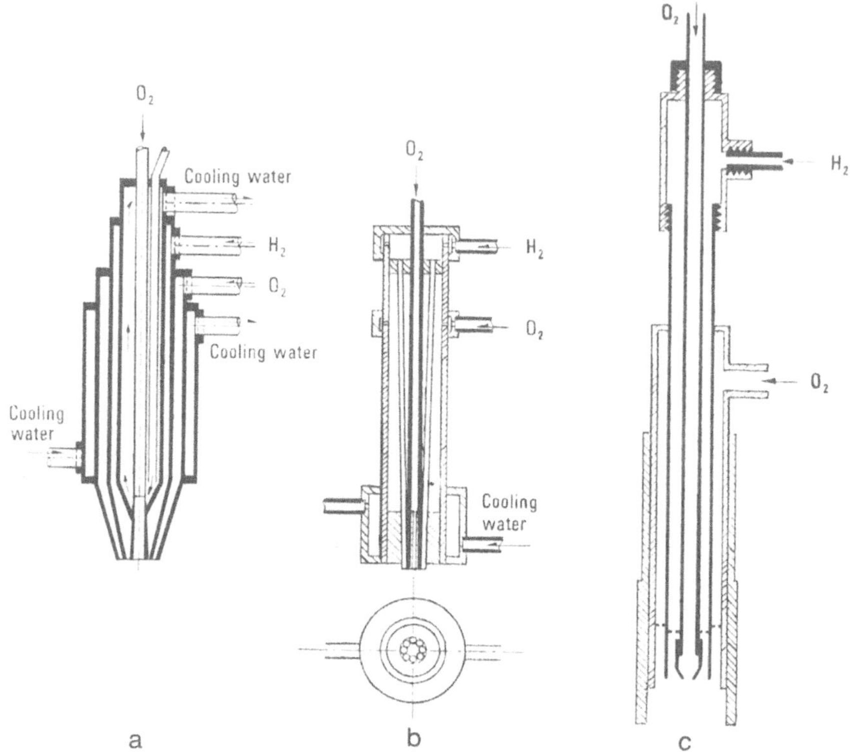

FIGURE 33. Three-tube burners after (a) Moore,[94] (b) Lefever,[93] and (c) Adamski.[92]

is carried by the outer tube. In multitube burners, i.e., consisting of more than two concentric tubes, the sequence alternates in a corresponding way. The gas exit is usually not by way of the interstices between the tubes, but through holes equally spaced on a circle. The larger the crystal and thus the burner, the more difficult it is to remove the heat transferred to the burner head from furnace, crystal, and flame. For crystals with a diameter of around 30 mm a simple water-cooling jacket in which the lower part of the burner is enclosed suffices (e.g., Fig. 33b). For larger diameters, however, not only has each tube to be cooled separately (see Fig. 33a) but, on the other hand, a certain closeness of spacing of the nozzle rings toward the outside is desirable—two requirements which cannot be satisfied simultaneously without some compromise.

It is probably because of the difficulty of cooling that no multitube burners are yet known in which the burner sections following the second concentric tube have been replaced by a large number of individual small burners. Such a construction should produce a more uniform flame front than that supplied by conventional multitube burners.

In Section 2 the importance of the heat loss due to radiation from the crystal top to the lower burner side has been mentioned. Burners with radiation reflectors made of high-melting rare metals like iridium or platinum may help in lowering these losses; however, no such work has been reported in the literature.

Detailed descriptions of burner constructions are to be found in the literature.[7,92,93] Some additional discussion is given in Section 9.

Burners are also known which differ from the version with parallel concentric tubes in that the tubes arranged around the central oxygen supply tube converge conically (Figs. 33a and 33b). The burner in Fig. 33a has been used to generate flames of high energy; it was developed by Moore[94] for the growth of particularly high melting oxides such as Y_2O_3, with a melting point of 2410°C.

In addition to supplying the gases the burner has the task of conveying the powder from the powder feed mechanism to the flame. For this purpose, the burner has a funnel-shaped extension at its upper end, through which the powder, dropping from a sieve, is passed to the central burner tube. The oxygen stream (typical gas velocity 500–1000 liters/hr cm^2), which is introduced above the funnel, entrains the powder. This entrainment effect must not be overestimated: even on well-polished walls it is found that powder occasionally piles up and subsequently falls like an avalanche when the piles have attained a certain size; this is particularly the case when the funnel–tube junction is sharp-edged and vortices form. The flame is not intense enough to melt such an avalanche completely before it reaches the melt film. This therefore leads to inclusions of unmelted material in the crystal. To prevent these pileups from occurring it has been suggested that the burner tube be vibrated by a small solenoid.[30] However, where a vibrator is used for powder conveyance, the residual vibrations transferred to the burner are usually sufficient to eliminate this undesirable effect.

With regard to the geometry of the burner, the length has to be such that the radiant heat, which enters the burner through the central (oxygen) nozzle, is dissipated and the powder feeder mechanism, particularly the sieve, is protected against excessive temperatures. With an inside diameter of the central tube of 10 to 12 mm a length of about 250 mm is adequate. The oxygen nozzle diameter depends on the diameter of the crystal to be grown. For crystal rods with a diameter of up to 5 to 10 mm, 2-mm holes are adequate, and 3 to 4 mm for diameters of 20 mm; 6 mm may be taken as the maximum value. In such a burner the pressure in front of the regulating valve lies usually between ten and a few hundred millimeters of water column. For the growth of sapphires or Mg–Al spinels with a diameter of about 15 mm the oxygen flow rate is around 500 liters/hr; about three times this amount of hydrogen is required. In multitube burners, the oxygen flow increases with increasing crystal diameter from the inner to the outer oxygen inlets. This is in order to get the hottest part toward the outside and help broaden the flame from what would otherwise be a pointed shape.

The distance between burner head and crystal is of great importance for growth, for the flame profile is changing as the distance from the burner increases.[8] Distance also depends on burner and furnace construction and on the crystal being grown, and must be determined experimentally for each setup; it is usually between 10 and 20 cm (see also Section 9).

6.3. Interaction between Flame, Growth Chamber, and Crystal

The interaction between burner and crystal as already noted is determined by burner construction, gas flow, and burner–crystal spacing. The interaction between crystal and furnace depends on the crystal diameter and also on the diameter of the growth chamber and its thermal

Table 2. Effect of Insulation, Burner Type, and Internal Diameter D of the Furnace Insert on the Maximum Crystal Diameter d_{max} That Can Be Achieved[a]

Arrangement				Crystal	
	Insert				
Insulation[b]	Type	Diameter (mm)	Burner[c]	d_{max}	Designation in Fig. 34
L	Massive	30	2-tube	20	a
L	Massive	40	2-tube	10	b
L + R	Massive	40	2-tube	30	c
R	Massive	50	2-tube	35	—
R	Massive	50	3-tube	40	—
R	Massive	60	3-tube	30	d
R	With adapter	40/60	3-tube	50	e

[a]Source: From Falckenberg.[7]
[b]L is poor, R is good. (Poor insulation means a heat conductivity of 0.8, and good 0.2 kcal/hr m^2 deg, measured at room temperature.)
[c]For dimensions of the burner, see Fig. 32. The furnace was as shown in Fig. 30; however, without lateral additional heating.

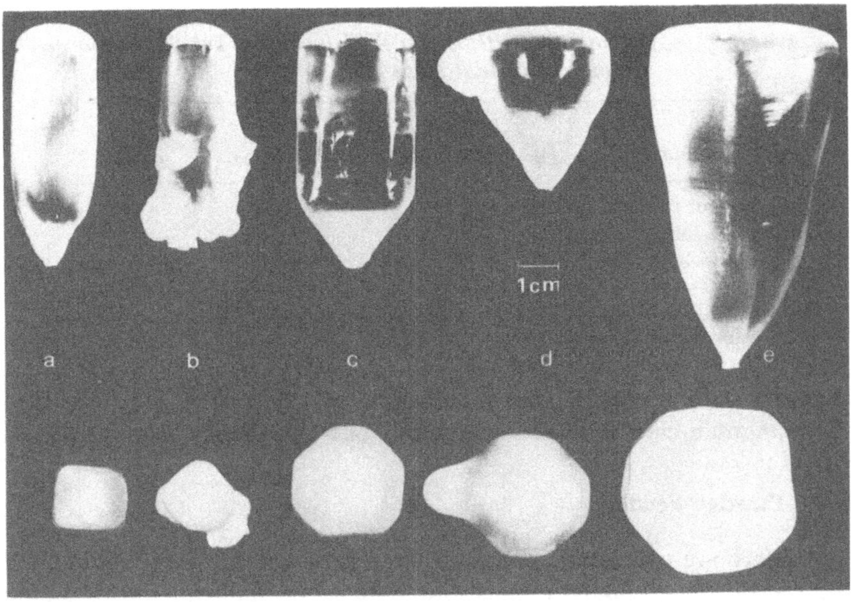

FIGURE 34. Mg–Al spinel crystals grown in furnaces of the type shown in Fig. 30, but without lateral burners. The series of crystals demonstrates the dependence of the crystal diameter on the parameters growth chamber and burner. For details see Table 2.

insulation. Detailed examination of the interactions has been undertaken for the furnace shown in Fig. 30, in which the flame gases are laterally deflected.[7] Table 2 shows the apparatus parameters and the crystal diameters that can be achieved under the specified conditions for Mg–Al spinel crystals with the composition $MgO:Al_2O_3 = 1:2$. The crystals obtained are shown in Fig. 34.

It can be seen from the table that as the diameter of the growth chamber increases, the unfavorable interaction between crystal and furnace makes it more difficult to obtain crystals with a large diameter. The increasing distance between the crystal and furnace walls leads to too steep a radial temperature gradient. Switching from a two-tube burner to a three-tube burner proves to be a necessary but not sufficient condition for a further increase in the diameter. The use of an adapter tube which acts as a heat shield proves to be necessary if the possibilities offered by the three-tube burner are to be fully utilized. The heat shields can be replaced by additional heaters as shown in Fig. 30.

Table 3 shows the maximum diameter of sapphire crystals that can be grown in well-insulated tubular furnaces. Only with a growth chamber diameter of 90 mm was the interaction between crystal and growth

Table 3. Maximum Achievable Diameter d_{max} of Sapphire Crystals (Orientation $\langle\bar{1}010\rangle$) in a Well-Insulated Tubular Furnace without Additional Heater[a]

Burner	Diameter of growth chamber	d_{max}	Aids
two-tube burner	40	35	—
three-tube burner	60	50	—
four-tube burner	90 and 70, respectively	80	adapter tube

[a]For diameters exceeding ~30 mm inadequate mechanical stability caused fractures within the crystals on cooling.

chamber so unfavorable that an adapter tube had to be used to achieve the maximum crystal diameter possible with this device, 80 mm (Fig. 35).

6.4. Powder Feeding

There are two sets of problems connected with powder feeding. First, the powder supply must replace the material crystallizing out of the melt film per unit time. This means that feeding must be continuous and precisely matched to the particular volume crystallizing out. Second, the powder must be supplied in a manner that does not give occasion for the creation of crystal defects, i.e., it should be finely divided and of low density. Let us start by discussing the latter set of problems.

6.4.1. Powder Transport

There are no data from experiments on the maximum permissible grain size. By virtue of the low thickness of the melt film, optimum results would be expected if the powder particles reach the melt film in the form of molten droplets. This possibility depends on the size and density of the powder particles and on other factors: namely, the distance between burner and liquid film, the wall temperature of the furnace tube, the gas flow rate, and the powder flow density. Without going into all the factors in detail, the progress of a particle through the apparatus will be outlined with the aid of Figs. 36–40.

Figure 36 shows sapphire powder (Al_2O_3 powder) obtained by heating alum. The noncompact nature of the powder is clearly evident. Analysis of a grain size distribution by sedimentation analysis is shown in Fig. 37.[95] Spinel powder made from alum looks similar in shape and size. Highly flocculent and low-density powder of the kind shown may generally be obtained by using starting materials which release gases like

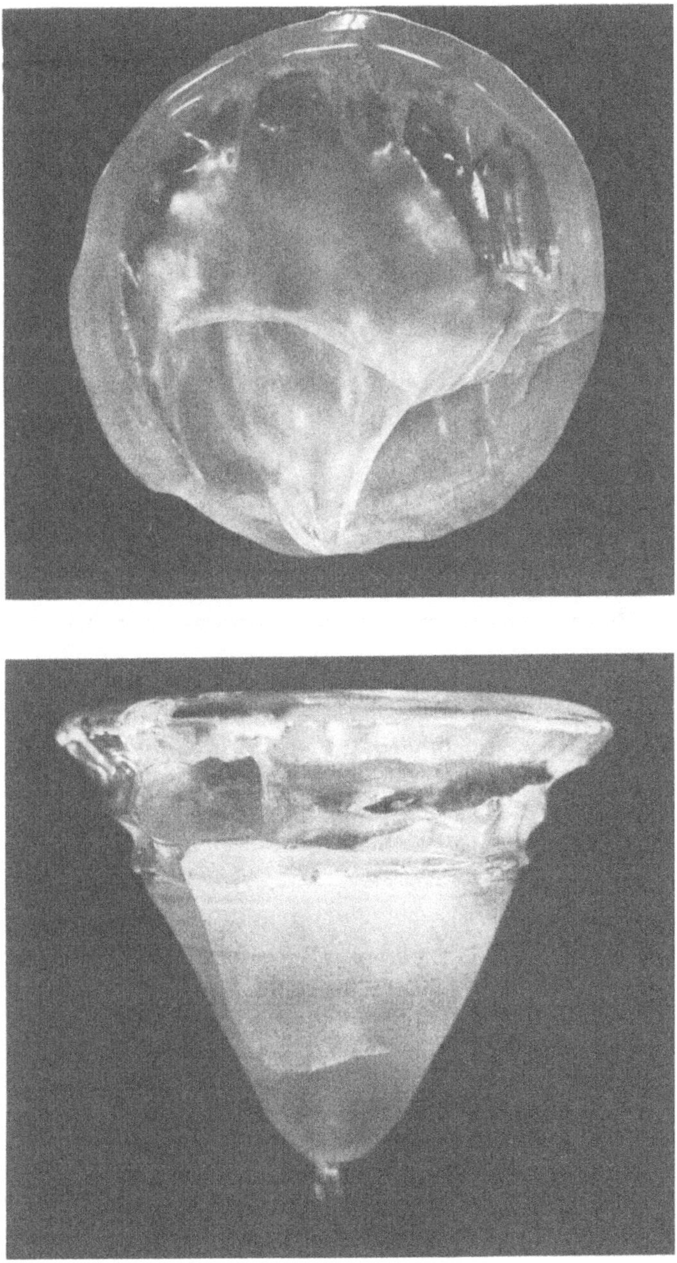

FIGURE 35. Two views of sapphire crystal with a diameter of 83 mm grown in a laboratory furnace with adapter tube. The cracks developed during cooling. Orientation $\langle\bar{1}010\rangle$.

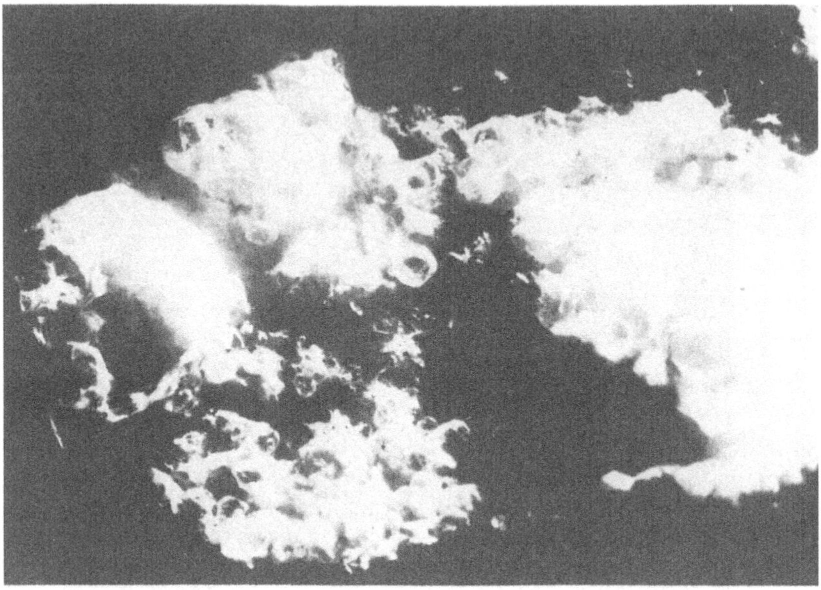

FIGURE 36. Sapphire powder obtained by calcination of alum. ×10,000.

NH_3, CO_2, H_2O, SO_2, or SO_3 during calcination. A comparison of Co_3O_4 powders made of starting materials of different compositions is given by Velikhova,[96] a survey of different materials was done by Wilke.[97]

From Fig. 37 it can be seen that the bulk of the powder grains have an effective diameter of less than 50 μm. The measured grain size

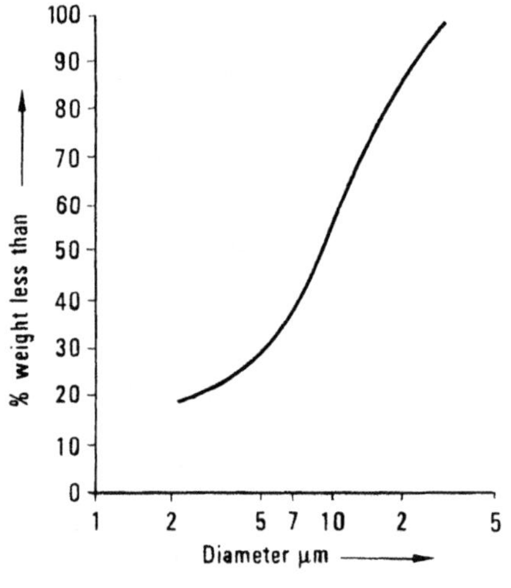

FIGURE 37. Grain size distribution for sapphire powder. (After Ricard.[95])

distribution[98] is, however, only of limited use for the assessment of the actual size of the grains passing through the sieve into the gas stream. This is because there is adhesion between the particles,[99] which exceeds the force of gravity for particle sizes below a diameter that is usually given as between 10 and 100 μm.[100] The forces of adhesion can be van der Waals attractive forces; they may also be forces of adhesion transferred via material bridges or forces of an electrostatic nature.

A scanning electron microscope photograph (Fig. 38) illustrates the effect of what are essentially van der Waals forces between glass beads.[100] However, there do not appear to have been any investigations of the powders used with the flame-fusion process; the remarks previously made are of quite a general nature and are intended to give an idea of the forces which may be active during powder transport. These forces manifest themselves in various ways when the powder is being fed. It is, for instance, impossible to sieve powder with the grain distribution shown in Fig. 37 through a sieve with a mesh aperture of 50 μm. Even with much outlay of mechanical energy 80 μm is about the lower limit for the mesh aperture of a sieve through which this material can still be passed. If little

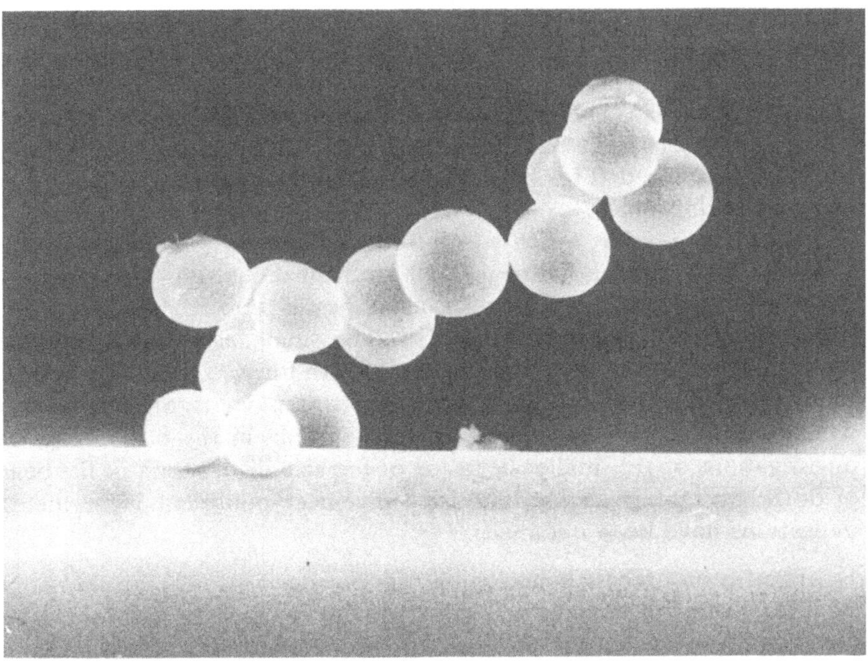

FIGURE 38. Glass spheres (diameter about 10 μm) adhering to a glass fiber surface. Scanning electron microscope photograph. (After Rumpf.[100])

FIGURE 39. Agglomeration of Mg–Al spinel powder below sieves with mesh aperture (a) 80 μm, (b) 50 μm.

energy is used, an agglomeration of filtered particles may be observed below the 80-μm sieve base (Fig. 39a); this effect increases as the mesh aperture is reduced (Fig. 39b). The sieve used for this investigation was made of Perlon fibers.

Sapphire powder, which was introduced into the gas stream through a sieve with a mesh aperture of 150 μm, can be used as an example to show what happens to a powder as it passes through the flame in a furnace with no crystal growing in it. The small fused beads produced were collected at the lower end of the growth furnace; Figs. 40a and 40b show polished sections of two beads. Both contain voids; the smaller bead contains fewer voids and these are smaller than in the other bead. No investigations of yet smaller beads or of the size distribution of the beads or of the percentage of fused particles have been published. Nevertheless, indications have been obtained:

1. The fusion process begins while the powder falls through the flame.
2. The reduction in volume associated with the melting is considerable; with a bulk density of the sapphire powder of 0.2 g/cm^3 and a sapphire density of 3.98 g/cm^3, the volume may be reduced by a factor of around 20.
3. The formation of voids decreases as the particle size is reduced.

Particles with a diameter of 80 μm would consequently be reduced in size to beads with a diameter of around 30 μm; voids would hardly be likely to occur then.

The origin of the voids is not known. It may be that gas is trapped as the highly irregular powder particles ball-up on melting; or there may be some kind of decomposition during melting, in the case of Al_2O_3 to form suboxide Al_2O; finally there may also be shrinkage during the cooling as the beads fall through the lower portion of the furnace. The latter, however, would not be of relevance for the growth process.

The period of stay of the particles in the melt film has so far been ignored. Taking into consideration the low thickness of the film, its influence should not be overestimated. At this point reference should be made to Section 3, which deals with questions concerning the melt film.

The other set of problems mentioned previously, associated with the replacement of the material crystallizing out per unit time during the growth process by a correspondingly large and constant powder stream, can now be discussed. The powder stream essentially has to satisfy three requirements. The first of these has already been explained: the grain size should not exceed a certain value, which we put as about 80 μm. A further requirement is for adequate throughput. The magnitude of the throughput depends on the crystal diameter, pulling speed, and loss of powder which does not reach the melt film because of scattering. The latter amounts to about 30%. The pulling speed is typically 1 cm/hr; for a sapphire crystal diameter of 20 mm the required throughput is then around 0.2 g/min, for 50 mm around 2 g/min.

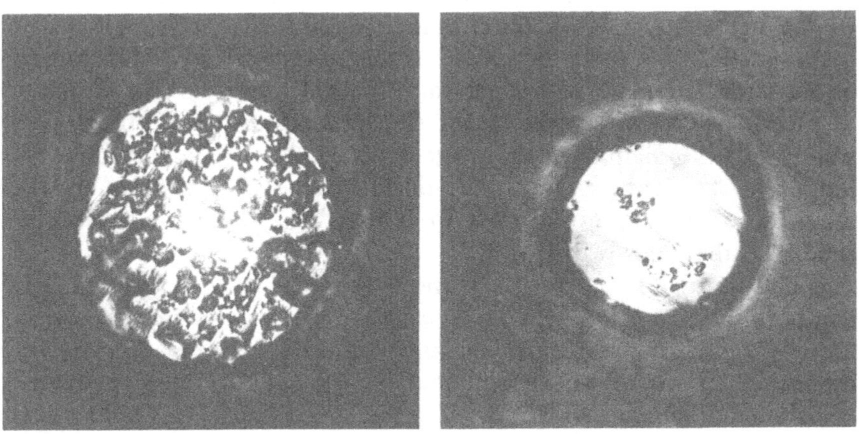

FIGURE 40. Sapphire powder particles after passing through sieve (mesh aperture 150 μm) and flame. ×500.

The final requirement concerns the short-term constancy (time scale 1 min) of the powder stream[101] which is more or less arbitrarily put at ±10%. This implies a ±10% variation in growth velocity. Many different kinds of powder feeder mechanisms have been developed[101]; they range from simple tapping mechanisms to the two-stage vibration feeder with combined horizontal and vertical motion for the growth of large crystals. The main problem in developing the feeder is the tendency of the powder to agglomerate.[68] This characteristic is particularly evident when a bed of powder is subjected to vibration under its own weight. The immediate result is powder densification with the consequence that the sieve becomes clogged and the powder stream "dries up." Vibrators are therefore normally used in a two-stage arrangement. The powder container is kept separate and works with its own simple feeder drive; it conveys powder to a second sieve, on which the powder depth is only very small, however. The pressure exerted by the powder's own weight is therefore also small, which has the significant advantage that this sieve can be subjected to the most varied types of vibration without becoming clogged up, as is otherwise unavoidable with a large amount of powder. The following section describes the various feeder mechanisms and compares their characteristics.

6.4.2. Single-Stage Feeder Mechanism

Among single-stage feeder mechanisms the most important has been the tapping mechanism, which in fact originated with Verneuil. This principle still finds widespread application today for the large-scale manufacture of gem and bearing stones on account of simplicity of construction and reliability of operation. The powder container, which contains at least as much powder as is required for one crystal, is closed at its lower end by a sieve. When either the elastically suspended container or the sieve is tapped with a hammer at regular intervals, small amounts of powder pass through the sieve. The throughput is determined by the weight of the hammer, the distance it falls, and its frequency of operation.[102] Figure 41 shows the throughput as a function of the mesh aperture for spinel powder made from alum. The grain size distribution is similar to that shown in Fig. 37.

The throughput decreases as the mesh aperture is reduced, independently of the hammer weight, and approaches zero at a mesh aperture of around 70 μm. The long-term constancy depends, among other things, on the depth of powder in the container. In the case of a mesh aperture of 80 μm and a hammer weight of 250 g the throughput drops over a period of 4 hr to about 40% of the initial value of 0.1 g/min (Fig. 42, curve a). For a larger mesh aperture the constancy is improved. Müller *et al.*[101]

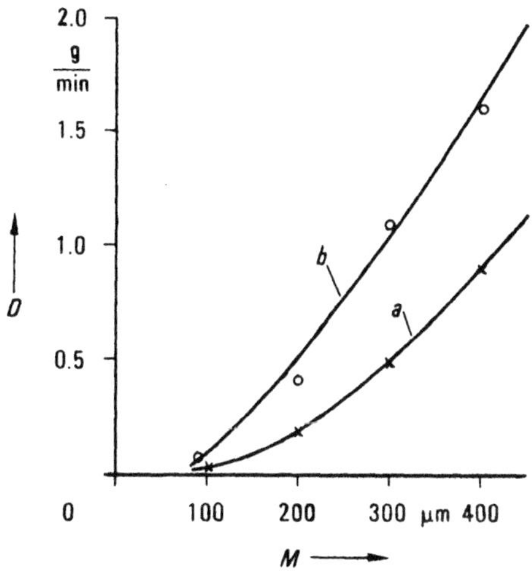

FIGURE 41. Powder throughput as a function of the mesh aperture M for a tapping mechanism; Mg–Al spinel powder. Weight of hammer (a) 90 g, (b) 250 g.

achieved a constancy within ±10% over a period of 3 hr for a mesh aperture of 150 μm.

The short-time constancy is naturally poor for the tapping process. As a result of the return motion of the sieve base, there is a brief suction effect following each hammer blow, which causes the powder flow to cease momentarily. The powder flow therefore fluctuates greatly even if the shock rate is high. However, it is doubtful whether these fluctuations,

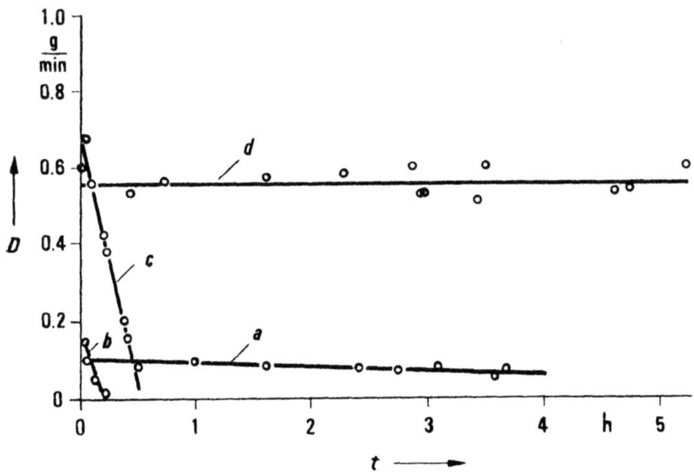

FIGURE 42. Powder throughput as a function of time t for a mesh aperture of 80 μm for Mg–Al spinel powder. a, Tapping mechanism; b, two-stage feeder mechanism with horizontal vibration of the lower sieve (a in Fig. 43); c, as b, but with Al_2O_3 spheres on the sieve; d, as c, but with additional vertical vibration.

typical frequency 100 min^{-1}, have any effect on the crystal quality. The frequency of striations in Verneuil crystals is, for example, much lower than the tapping frequency (see Section 3).[17]

The tapping mechanism can thus be characterized as follows: The long-term constancy is relatively good if the mesh aperture is not too small; the critical mesh aperture, below which throughput "dries up," is about 80 μm for typical sapphire or spinel powder obtained from alum. As a result of the agglomeration tendency of the powder this limit cannot be reduced appreciably by the use of a preliminary fine-sieving process. At this mesh aperture the throughput is not high enough to allow the growth of crystals with a diameter larger than about 15 mm.

Equipment has also been developed in which the hammer mechanism is replaced by an electromagnetic vibrator.[103] However, this has the disadvantage that the sieve clogs up quickly or the throughput is irregular.[68,101]

Further experiments to obtain continuous powder feeding have been motivated by the wish to improve crystal perfection and to increase the powder throughput at small mesh apertures.

6.4.3. Two-Stage Feeder Mechanism

With the tapping mechanism the powder drops directly into the burner tube from a powder container: With two-stage devices an additional stage is provided between container and burner. A simple two-stage device using a feeder trough without sieve has been developed by Carter[104]; the constancy of the powder flow as measured by Müller *et al.*[101] is ~30%.

Three types of two-stage mechanisms with sieves have been developed which are distinguished by the manner in which the vibration is imparted to the fine dosing sieve. In the feeder mechanism developed by Seifert[68] the sieve vibrates in the horizontal direction; a constant powder throughput is achieved with a mesh aperture of above ~250 μm. In the mechanism developed by Müller *et al.*[101] an oscillating magnet acts in a vertical direction on a sieve with asymmetrical spring retention, so that a vibratory movement about a horizontal axis is superimposed on the vertical movement; the constancy of the throughput was examined for a mesh aperture of 150 μm.[101] In the following a mechanism is described in which a purely vertical vibratory movement is superimposed on a purely horizontal vibratory movement[107] (Fig. 43). With this setup constant throughputs can be achieved for mesh apertures of 80 μm.

In the two-stage mechanism shown in Fig. 43, the fine dosing sieve (a) is held by two spring steel sheets (b) and coupled to the oscillating

FIGURE 43. Schematic diagram of a two-stage powder feeder mechanism with horizontal and vertical excitation of the fine dosing sieve.[107] a, Fine dosing sieve; b, spring steel sheet; c, armature; d, vibration magnet; e, light barrier; f, springs acting in a vertical direction; g, vibration magnets working in vertical direction; h, hammer acting in horizontal direction; i, coarse dosing sieve; j, powder container; k, pressure compensating tube.

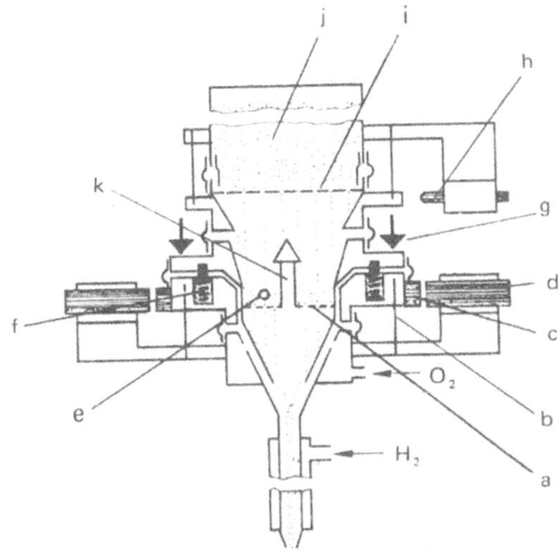

magnets (d) by means of two armatures (c). One of the two oscillating magnets is driven by the upper, the other by the lower half-wave of the power network voltage. Two oscillating magnets working in the vertical direction (g) also act on the fine dosing sieve (a). The springs (f) counteract the magnets (g). The height of the powder on the fine dosing sieve (a) is kept constant with the aid of the light barrier (light source and photocell, e), which controls via a relay the powder flow from the powder container (j) to the fine dosing sieve (a) by actuating the hammer (h), which acts on the coarse dosing sieve (i). The throughput through the fine dosing sieve (a) can be regulated by altering the amplitude of the oscillations.

The operating principle is first considered for a purely horizontal oscillation; the vertical oscillation is blocked. Figure 42, curves b and c, shows the dependence of the throughput on time, using a sieve with a mesh aperture of 80 μm. The spinel powder was the same as used for Fig. 39a. The throughput decreases rapidly, the overall rate falling off about 10 times faster than is the case with the tapping mechanism. The introduction of aluminum oxide spheres, diameter 4 mm, increases the initial throughput considerably (Fig. 42, curve c), but does not alter the rate at which the throughput decreases. It is not until a mesh aperture of ~250 μm is used that a constant powder throughput is achieved over an adequately long period without employing Al_2O_3 spheres.

On applying the vertical oscillations with a frequency of 100 Hz and introducing a layer of Al_2O_3 spheres, the result is as follows with a sieve of 80-μm mesh aperture in the fine dosing stage: When the light barrier is used and the powder throughput is regulated by altering the amplitude of the oscillating magnets, the sieve clogs up in a short period. If, on the other hand, the coarse dosing sieve with a mesh aperture of around 800 μm is used to regulate the throughput, and if the horizontal and vertical working oscillating magnets of the fine dosing sieve are allowed to work at full power, very even throughputs are obtained up to amounts of several grams per minute (Fig. 42d).

The function of the alumina spheres is to enhance the effect of vibration. Kinetic energy is transferred to them by the vibrating sieve; by milling and thrusting motion they contribute to passing the powder through the sieve.

Mention should be made here of two devices developed to grow crystals of varying composition. Adamski[105] used a carrier gas-diverting apparatus, Sampson[106] a double-chamber powder container.

6.4.4. Aerodynamic Effects

When powder feeder mechanisms are used, a number of aerodynamic effects occur as a result of interactions between sieve movement, gas stream, and powder stream.

Seifert[68] mentioned a pressure compensating tube which was attached to the fine dosing sieve to obtain an even powder stream (Fig. 43). The oxygen, which flows past the opening of the funnel arranged below the fine dosing sieve (a), lowers the pressure within the funnel. The air column lying above the sieve therefore exerts a pressure on the powder, with the result that the throughput is reduced. The difference in pressure between the spaces above and below the sieve is balanced by the tube, and depending on the geometry of the funnel, the throughput can be increased two- to threefold.

Another disturbing effect is observed when the vibration of the fine dosing sieve has a vertical component. These oscillations are passed on by the air column and influence the throughput through the coarse dosing sieve (Fig. 43, i). The throughput can no longer be controlled with the required precision. A "conveyer sphere" between coarse and fine dosing sieve reduces the cross-sectional area to a small gap and stops the interaction.[107] An even powder flow through the gap is obtained by rotating the sphere about a horizontal axis.

7. REDOX NATURE OF THE H_2/O_2 FLAME AND VOLATILIZATION

During the growth process powder particles and melt are heated by the H_2/O_2 flame gases. The temperature and chemical composition of the flame gases influence not only the heat exchange but also the physicochemical processes in powder and melt. A thermodynamic analysis of composition and temperature of the combustion products carried out by Bazhenova *et al.*[108] covered the range of H_2/O_2 composition ratios of 0.8–8.0. In order to obtain an approximate estimation of the redox potential of the combustion products of H_2/O_2 relative to some particular oxide the authors used the partial pressures in the thermal dissociation products of the oxides. By comparison with the partial pressures of the flame gas components: H_2, O_2, H_2O, H, O, and OH, it is shown, for example, that for oxides like Y_2O_3, ZrO_2, ThO_2, Al_2O_3, La_2O_3, and Nd_2O_3 the atmosphere is oxidizing up to compositions $H_2/O_2 \approx 3:1$. For these oxides one does not have to worry that the stoichiometry of the crystal will be disturbed or that it will volatilize in the flame owing to reduction. In those cases in which the gas atmosphere is reducing, the authors suggest the admixture of small amounts of an inert gas in order to reduce the concentration of the reducing agent. If one does not consider redox processes, the growth conditions are more favorable when there is excess hydrogen in the flame: combustion temperature and heat emission are somewhat higher with H_2 excess than with O_2 excess; in addition the excess hydrogen burns at the outlet and thus reduces the heat loss of the furnace.

Experimental investigations of the influence of the flame composition on the chemical composition of various metal oxide crystals have been made by Popova *et al.*[109–112] The valence state of the crystals is essentially defined by the composition of the flame and not of the starting material. From Mn_3O_4 powder, crystals of either MnO or Mn_3O_4 could be grown depending on the H_2/O_2 ratio (see Table 4). Similar results have been obtained for various Fe oxides. The behavior of Fe and Ni in Al_2O_3[110] as well as of the rare earth oxides[111] has also been studied in this context.

The Cr_2O_3 content in ruby has been investigated by Pastor *et al.*[113] Experiments with the H_2–O_2 and the O_2–town gas flame and with differently prepared powder produced no difference in the rate of chromium evaporation. The powder particles evidently stay in the flame for too short a time for the redox properties of the flame to come into effect. On the other hand, a distinct relationship was found between the rate of evaporation and the distance of the crystal from the burner.[114] By varying this distance it is possible to influence the distribution of the chromium in the ruby. If the oxygen cone of the flame is in contact with the melt on the crystal, the crystal grows with a low chromium content in the region of the longitudinal axis.[16,115]

The evaporation of chromium was measured by Sil'nichenko and Gritsenko.[116] According to these measurements the ratio of chromium in the crystal to chromium in the original powder is 0.3. The measurements

Table 4. Chemical Stability of Oxides and Flame Composition during the Growth Process[a]

Composition	$H_2:O_2$
MnO	4.0–4.1 : 1
Fe_3O_4	2.5–2.6 : 1
NiO	2.1–2.15 : 1
CoO	1.9–2.0 : 1
Mn_3O_4	1.55 : 1

[a]After Popova.[109]

were made with up to 1% chromium content in the crystal. A transition from Cr^{3+} to Cr^{4+} in the ruby was achieved by adding nitrides to the powder.[117]

8. MECHANICAL STRENGTH AND ANNEALING

The stresses existing in crystals after growth often lead to cracking during machining. In order to reduce or eliminate the stresses,[118] the crystals may be annealed. This must be performed at such a high temperature that under the action of the stresses a plastic deformation takes place to relieve them. Subsequent cooling must be so slow and even that no new stresses can develop. In some cases, e.g., with Mg–Al spinels, it is possible to obtain extra strength through precipitation strengthening. In such cases, the temperature at which the heat treatment takes place is lower than for annealing, as is discussed later.

Annealing is characterized by the temperature program to which the crystal is subjected, by the type of furnace used, and by the gas atmosphere in which annealing takes place. The choice of these parameters depends on the material being annealed. A successfully tested temperature program for rubies with the dimensions 130 × 15 mm is shown in Fig. 44.[119] In the first phase (A), the crystals are heated to the annealing temperature; this is followed by phase B in which they are annealed at a temperature close to the melting point, and phase C in which they are cooled to room temperature.

For the annealing tests, a high-temperature, high-vacuum annealing furnace with a molybdenum cylinder as heater was used; the rubies were suspended from a tantalum holder. The background pressure was 10^{-4} to 10^{-5} Torr. Strength tests were based on the reaction of the annealed rubies to mechanical processes such as cutting and polishing.

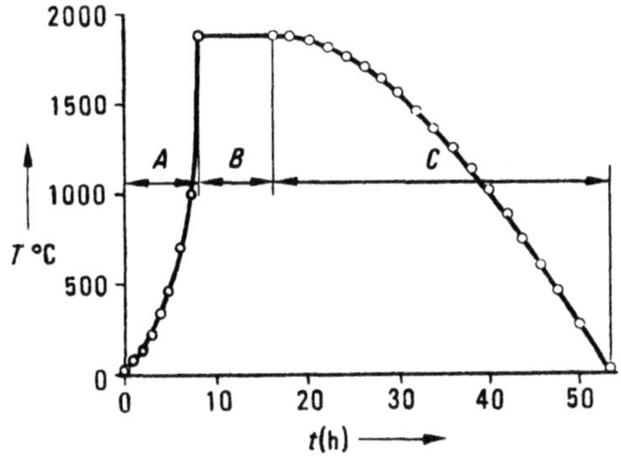

FIGURE 44. Annealing curve for ruby crystals with a length of 130 mm and a diameter of 15 mm. (After Müller.[119])

The duration of phase B depends to a very great extent on the temperature. Figure 45 shows that although plastic deformation of the ruby is observed even at such relatively low temperatures as 1600°C, adequate stress relieving takes 800 hr.[119]

Kvapil *et al.*[120] have established an exponential relationship between strength and annealing time for corundum. Kvapil determined the magnitude of the stresses by the stress birefringence method by viewing laterally in the direction of the *c* axis, which formed an angle of 60 to 80° with the growth axis. For the residual stresses T_K after annealing for a duration t one can state

$$T_K = T_{K_0} \cdot e^{-Kt}$$

The time constant K rises with increasing temperature and falls with increasing crystal diameter, and, in the case of ruby, also with increasing chromium content.

The influence of the atmosphere in the furnace has scarcely been investigated. Experimental results are only available for ruby. It was found that rubies are eroded when annealed in open graphite susceptors,

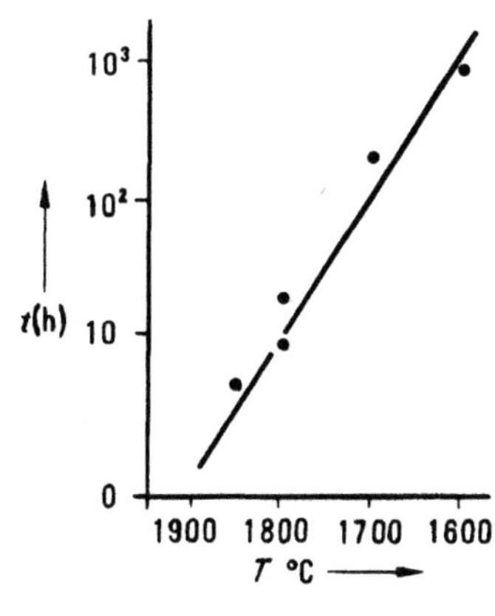

FIGURE 45. Dependence of the annealing time t on temperature T. (After Müller.[119])

but that erosion can be prevented by using a CO atmosphere.[64] Probably, C reduces Al_2O_3 to give Al_2O, which is volatile. In some cases, it is sufficient if the graphite susceptor is tightly closed, so that enough CO pressure can build up. Annealing in nitrogen results in a gray bloom due to a high-temperature reaction of nitrogen with the aluminum oxide, thereby forming aluminum nitride. The extent of the reaction depends on the chromium content of the ruby. The effects of various media such as CO, H_2, vacuum, and O_2 during heat treatment on the optical properties of ruby have been investigated by Sandreyev *et al.*[121] It should be noted that annealing can remove stresses, but not dislocations or grain boundaries. At most, these will be rearranged.[50,122,123]

In some cases it is possible for the crystals to be strengthened by the formation of the preliminary stages of precipitation. Figure 46 shows the influence of the cooling process in tubular crystal-growing furnaces with different thermal capacities on the strength of Al_2O_3-rich Mg–Al spinels.[124] It is characteristic of this type of strengthening[125] that annealing at 1800°C followed by rapid cooling results in lower mechanical stability than annealing at 850°C. The reason for this behavior is the formation of a strengthening phase (Guinier–Preston zones) which is transformed to other phases as the temperature is increased; at high temperatures a homogeneous solution is finally formed.[126] This kind of behavior is well known in alloy metallurgy.[127]

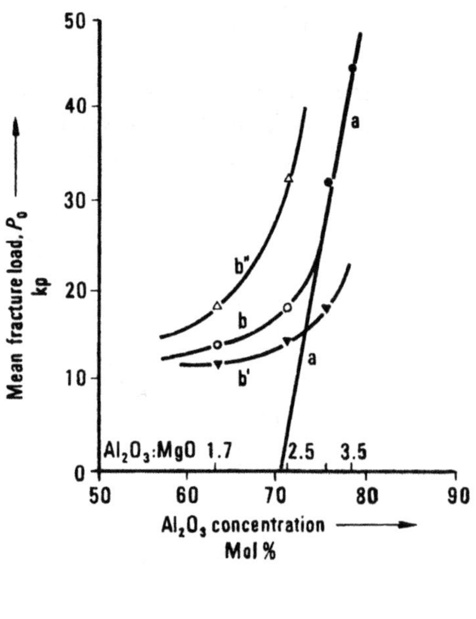

FIGURE 46. Precipitation strengthening of Mg–Al spinel crystals. a, Crystals grown in a furnace with low thermal capacity, unannealed; b, crystals grown in a furnace with high thermal capacity, unannealed; b′, crystals annealed at 1800°C; b″, crystals annealed at 850°C.[124]

9. UNUSUAL MATERIALS GROWN BY THE VERNEUIL METHOD AND SPECIAL TECHNIQUES

There are comprehensive indexes and references in the literature regarding crystals grown by the Verneuil method.(37,97,128–130) In the following, a collection of more recent papers is given. Some of the references have been selected in order to point out special techniques or modifications to apparatus or methods. It is difficult to assess the general value of the modifications, since apparatus differences are great and there is no basis for comparison. Some modifications may rectify a disturbance in certain cases which in other cases does not even occur, because of differences in the respective apparatuses.

Homogeneous single crystals of mullite with the composition $2Al_2O_3 \cdot SiO_2$ were grown by Neuhaus and Richartz.(131) Manufacture was made possible by the use of a water-cooled crystal carrier with high, controllable cooling penetrating into the crystal. The crystals had a length of up to 2.5 cm and a diameter of up to 0.8 cm; the melting temperature lay between 1850 and 1950°C. Mullite crystals were also grown by Bauer and co-workers(132,133) and by Barta and Barta.(134)

Single crystals of lanthanide oxides (La_2O_3 and Nd_2O_3 with hexagonal structure, Sm_2O_3 and Gd_2O_3 with monoclinic structure, Er_2O_3 with cubic structure) were grown by Lejus *et al.*(135,136) The crystals had a length of up to ~30 mm and a diameter of up to ~8 mm. A H_2/O_2 flame was used. Since the melting points of these particular materials lie between 2400 and 2500°C, stabilized zirconium oxide rods were used as seed holders. The furnace used was of muffle type with two half-shells, composed chiefly of Al_2O_3, with internal diameter of 35 to 40 mm, and an external diameter of 130 mm.

Single crystals of forsterite, Mg_2SiO_4, melting point 1898°C, were grown by Shankland.(30) The necessity for a polished inner surface of the feed funnel was avoided by inserting a lining of hard paper, which was made to vibrate by means of an electromagnetic device. In order to prevent cracking of the crystals, a Pt–13% Rh heater was installed in the growth furnace. A three-tube burner with conical nozzle was used. The crystals were up to 5 cm long with a diameter of about 10 mm. In order to prevent cracking during machining, it was necessary to anneal them at 1750 to 1800°C. A series of silicates, including forsterite, were also grown by Bauer and Gordon.(137) Silicates of the rare earths have been grown by Bondar' *et al.*,(138) and of calcium by Saalfeld.(139,140) Crystals of chromium oxide, Cr_2O_3, melting point ~2275°C, were grown by various authors.(141,142) A detailed description is given by Lefever and Matsko.(143) These authors used a three-tube burner(144) and a furnace of Al_2O_3 bricks with an inspection window covered by quartz glass; the flame gases were exhausted at the bottom on one side. The crystals could be cooled by gradually diminishing the flame; annealing was not necessary. The crystals were 7 cm or more in length and 1.5 cm in diameter; the pulling speed was 0.5 mm/min.

Mg_2TiO_4 crystals were grown by Dittmann and Petzelt.[145] This material melts incongruently at 1840°C. One difficulty in growing these crystals is that a phase change takes place between 1300 and 1500°C which can only be prevented by fast cooling in this temperature range. The crystals obtained therefore were cracked. In order to keep defects in the oxygen lattice to a minimum, a three-tube burner was used, which ensured a surplus of oxygen. The pulling speed was 2 to 3 mm/hr. The small broadening angle is ascribed to low surface tension.

Rare earth oxides with a cubic crystal lattice (Y_2O_3, Er_2O_3, Yb_2O_3) were grown by Lefever in such a way as to remain free from cracks.[146] In a Verneuil apparatus the flow of powder is known to spread out in a scatter cone below the oxygen nozzle. Lefever found that, with a suitable ceramic base positioned at about the same height as the seed crystal, it was possible to make a thick porous layer of sintered powder form around the crystal as it grew (Fig. 47) which provided good thermal

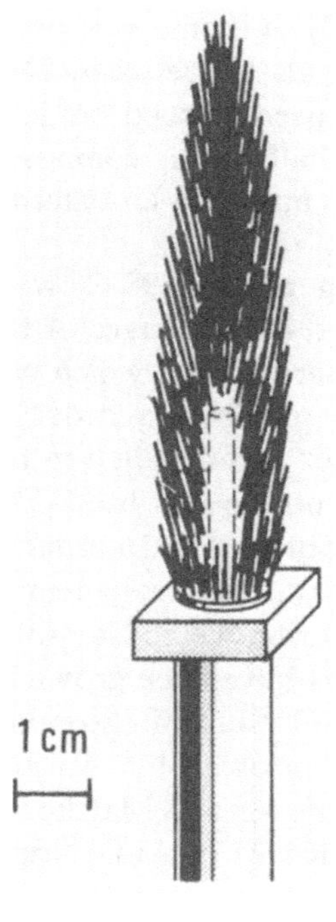

FIGURE 47. Formation of thermally insulating sheath by sintering powder around a growing crystal. (After Lefever.[146])

insulation on cooling. When this was ground off, Lefever obtained crystals with a diameter of 3 mm and a length of 20 mm that were mechanically stable. In order to reach the high melting temperature (about 2410°C for Y_2O_3), a three-tube burner with conical nozzle was used.[93] A method of growing these crystals developed by Pastor has already been described in Section 5.

Adamski *et al.* grew rubies whose *c*-axis was parallel to the growth axis.[18] The furnace used was based on a cast Al_2O_3 muffle inside which two Al_2O_3 tubes, one within the other, served as heat reflectors. The inner tube could be easily exchanged in order to prevent contamination when growing other types of crystals. The entire apparatus was housed in a steel cabinet, thus avoiding air contamination and simplifying supervision of the flame configuration. The grown rubies had hexagonal morphology—in contrast to the usual 0° rubies, which grow with a round cross section.

The behavior of binary systems of rare earth oxides during crystallization by the Verneuil method was investigated by Popova.[147] Using a two-tube burner, hydrogen and oxygen were combined in a ratio of about 1:1; the absolute gas volume varied slightly depending on the melting point of the oxides. On termination of the growth process, the crystals were taken out of the furnace usually within 3 to 4 min. Changes in cooling procedure had no detectable effect on the phase composition or, perhaps surprisingly, on crystal quality.

Ruby crystals were grown by Shiroki[69] using an automatic control mechanism in order to avoid crystal defects that arise due to variation in the distance between the burner and the growth front. The mechanism made use of the fact that the furnace temperature in the vicinity of the molten crystal top is essentially determined, because of the low heat emission of the flame, by radiation from the crystal. A Pt/Pt–13% Rh thermocouple was therefore mounted in the furnace wall above the level of the growth front. This controlled the pulling motor in such a way that the height of the molten top remained constant within ±0.5 mm. The sensitivity of the control mechanism depended to a great extent on the crystal diameter. At the end of the broadening phase the inspection hole was sealed. Another paper by Shiroki describing the growth of rubies with the aid of additional side burners has already been mentioned in Section 5.2.[15]

Rutile (TiO_2), strontium titanate ($SrTiO_3$), nickel oxide, sapphire, ruby, and spinel ($MgO:Al_2O_3 = 1:3$) with diameters of 24 mm and greater were grown by Adamski with a three-tube burner.[92] The burner essentially consisted of three concentric tubes (Fig. 33c) so as to avoid the sharply defined, turbulent mixing zones that occur with burners having conical nozzles. TiO_2 has also been grown by Kestigian.[148]

Stoichiometric Al–Mg spinels have been grown by a number of authors. The crystals crack very easily in comparison with spinels rich in Al_2O_3 ($MgO:Al_2O_3 = 1:3.5$).[149–152] Frequent use was therefore made of furnaces with afterheaters[70,71] or with special structure,[31] as already described in Section 6. Wang[153] used a three-tube burner based on the design of Adamski[92] and also a furnace that was fully enclosed in a container during use, in order to prevent flame fluctuations. The crystal produced was 20-mm long and about 10 mm in diameter.

Yamamoto *et al.*[154] grew sapphire with an oxygen/town gas jet, the gas ratio being 13:10 (liters/min); the crystals were ~10 mm in diameter.

An attempt was made by Haberland[155] to grow MgO crystals. Because of the high melting point of 2800°C, an oxyacetylene flame was used (see Table 1). The flame propagation rate of acetylene is high, namely, 1350 cm/sec (compared with hydrogen, 890 cm/sec); in order to keep the flame burning, the gas must have a high enough velocity. The result of this was that during the growing process the droplets of melt were blown past the seed carrier. It was likewise possible to melt ZrO_2 and BeO, but due to the high gas pressure it was only possible to produce fir-tree-shaped columns of fused material.

According to Kvapil[157] the mechanical stability of single corundum crystals depends on the ratio of the additive concentration in the single crystal to that in its seed crystal. With increasing concentration of the usual additives, e.g., Cr^{3+}, in the single crystal, the stability increases, but it decreases as the concentration of this additive in the seed crystal increases.

Scattering centers in crystals can be detected by transmitting a laser beam through them.[158] Adamski *et al.* found that among the rubies grown in the same apparatus, but having different orientation, those with 0° orientation (c axis$\parallel$growth axis) have the fewest scattering centers.[18]

Corundum single crystals of large diameter have been grown by Ricard and Cioccolani,[159] Zemlicka,[23] Belyaev *et al.*,[72] Barta,[160] and Falckenberg.[60] A number of authors report on the quality of ruby crystals[161,162] as found by optical[163,164] and etch[165] investigations. Dislocations have been investigated by etching techniques for sapphire crystals by Kliya,[166] Stephens and Alford,[167] and Palmour *et al.*,[168] and for Mg–Al spinel by Grabmaier and Watson.[169] Single crystals of scandium oxide,[170] $CaWO_4$ and $CdWO_4$,[171] cobalt oxide,[96] and other transition element oxides[172–174] of (Mg, Mn)O–Al_2O_3 and Fe_3O_4,[174] and $SrTiO_3$[175] have also been grown. The mosaic structure in ruby–sapphire junctions has been investigated by Kikin.[176] The crystallization process near the melting point was investigated for corundum by Dräger and Studt.[177]

10. CONCLUDING REMARKS

In the last few years it has been found, taking spinel and sapphire as examples, that it is possible to grow large crystals with reasonable mechanical stability by the Verneuil process. If one assumes flame compatibility, the main drawback of the method is that so far it has not proved possible to grow crystals of such perfection as can be obtained by the Czochralski technique. This chapter supports the view that poor crystal quality is not inherent in the method. Assuming efficient mechanisms for powder feed and crystal pulling, and precise control of the gas jet, the only reasons for poor crystal quality can be excessive grain size or powder density, and over-rapid cooling of the crystal. Since sapphire powder, for example, can today be produced with a density of only one-twentieth of that of the solid crystal and since a grain size of 80 μm is possible, only droplets of melt ought to reach the melt surface. There would thus no longer be any direct reason for defect formation from this side. All that remains is cooling and it is that which presents the greatest problems. As mentioned in Section 1, the direction of energy supply, which is parallel to the longitudinal crystal axis and not perpendicular to it, is unfavorable in comparison with other growing methods. The investigations with sapphire and spinel clearly show that axial and radial temperature gradients lead to plastic deformation and fracture. The axial gradient in the growth chamber determines the dependence of the radial gradient in the crystal on the length. It also determines the axial gradient in the crystal. The strength and perfection of the crystals thus depend on the temperature and its distribution on the flank of the growing or cooling crystal. Not until this temperature can be controlled will it be possible to investigate whether and to what extent other factors, such as the method of material feed, still affect the perfection of the crystals. With increasing wall temperature, the question of the stability of the melt film arises; its instability sets a final limit. The question arises of how the crystal temperature can be increased in practice. Direct and indirect heating methods have been quoted; progress has been made in the past few years in putting these methods into practice, but this still remains the subject of future work.

As to the problem of flame compatibility, the plasma torch offers the possibility of using neutral or reducing atmospheres. This approach might open up Verneuil-type growth of, say, transition metal carbides or silicides, or even of molybdenum and tungsten metals themselves.

In summary, the Verneuil process, probably the oldest method of growing single crystals on an industrial scale, is still far from fully developed. Research into its improvement and into the broadening of its range of applications is still highly active, and there is rich promise of future developments.

REFERENCES

1. A. Verneuil, *Ann. Chim. Phys.* **3**, 20 (1904); *Compt. Rend.* **135**, 791 (1902).
2. K. Nassau, *J. Crystal Growth* **13/14**, 12 (1972).
3. T. B. Reed, *J. Appl. Phys.* **32**, 2534 (1961); M. v. Ardenne, E. D. Knebel, H. Wachtel, and P. Wiese, *Krist. Techn.* **1**, 437 (1966); W. J. Alford and W. H. Bauer, *J. Phys. Chem. Solids* Suppl. No. 1, 71 (1967); D. T. Williams and W. A. Smith, *J. Am. Ceram. Soc.* **51**, 32 (1968); D. R. Kinloch and C. E. Birchenall, *J. Crystal Growth* **19**, 105 (1973).
4. R. E. De La Rue and F. A. Halden, *Rev. Sci. Instrum.* **31**, 35 (1960); F. A. Halden and R. Sedlacek, *Rev. Sci. Instrum.* **34**, 622 (1963); R. W. Bartlett, F. A. Halden, and J. W. Fowler, *Rev. Sci. Instrum.* **38**, 1313 (1967).
5. M. J. Musatov, *Sov. J. Opt. Technol.* **42**, 461 (1975).
6. J. Ricard, IEEE SOS Workshop, Lake Tahoe, Calfornia, September 1975 (unpublished).
7. R. Falckenberg, *J. Crystal Growth* **29**, 195 (1975).
8. F. B. Khambatta, P. J. Gielisse, M. P. Wilson, J. A. Adamski, and C. Sahagian, *J. Crystal Growth* **13/14**, 710 (1972).
9. V. A. Shchelkotunov and V. N. Danilov, *Inorg. Materials* **5**, 517 (1969).
10. E. E. Shpilrain, K. A. Yakimovich, and A. F. Tsitarkin, *High Temp.–High Press.* **5**, 191 (1973).
11. L. S. Barkhatov, D. N. Kagan, A. F. Tsitsarkin, E. E. Shpilrain, and K. A. Yakimovich, *High Temp.–High Press.* **11**, 1063 (1973).
12. J. J. Rasmussen and R. P. Nelson, *J. Am. Ceram. Soc.* **54**, 398 (1971).
13. D. D. Eley, ed., *Adhesion*, Oxford University Press, London (1961).
14. F. Ordway and P. R. Miller, *Am. Ceram. Soc. Bull.* **43**, 253 (1964).
15. K. Shiroki, *Japan J. Appl. Phys.* **8**, 1082 (1969).
16. J. Kvapil, B. Perner, and J. Kvapil, *Krist. Techn.* **9**, 503 (1974).
17. J. G. Grabmaier, *J. Crystal Growth* **5**, 105 (1969).
18. J. A. Adamski, R. C. Powell, and J. L. Sampson, *J. Crystal Growth* **3/4**, 246 (1968).
19. R. Falckenberg and H. Wörl, *Materials Res. Bull.* **9**, 519 (1974).
20. A. Neuhaus and K. Brenner, *Chem.-Ing. Technol.* **27**, 320 (1955).
21. A. Neuhaus, *Fortschr. Min.* **34**, 35 (1956).
22. J. N. Akimovich, *Ukrain. Fiz. Zh.* **14**, 1197 (1969).
23. J. Zemlicka, *Ber. Geol. Ges. DDR*, **7**, 492 (1962).
24. H. S. Bagdasarov, G. V. Berezhkova, V. G. Gorovkov, and A. E. Smirnov, *J. Crystal Growth* **22**, 61 (1974).
25. A. Müller and M. Wilhelm, *Z. Naturforsch.* **19a**, 254 (1964).
26. D. T. J. Hurle, *J. Crystal Growth* **13/14**, 39 (1972).
27. H. Pink, *Phys. Status Solidi* **21**, K111 (1967).
28. H. Wörl and R. Falckenberg, *Materials Res. Bull.* **11**, 807 (1976).
29. M. I. Musatov, *Sov. J. Opt. Technol.* **41**, 217 (1974).
30. T. H. Shankland, *Am. Ceram. Soc. Bull.* **46**, 1160 (1967).
31. R. Falckenberg, *J. Crystal Growth* **13/14**, 723 (1972).
32. J. Kvapil, *Krist. Techn.* **4**, 117 (1969).

33. V. L. Indenbom and G. E. Tomilovskii, *Kristallografiya* **3**, 593 (1958).
34. B. Lewis, R. N. Pease, and H. S. Taylor, *High-Speed Aerodynamics and Jet Propulsion Combustion Processes*, University Press, Princeton, New Jersey (1956).
35. W. Heywang and G. Ziegler, *Z. Naturforsch.* **9a**, 561 (1954).
36. W. Heywang, *Z. Naturforsch.* **11a**, 238 (1956).
37. R. A. Laudise, *The Growth of Single Crystals*, Prentice-Hall, Englewood Cliffs, New Jersey (1970).
38. S. V. Tsivinsky, *Krist. Techn.* **10**, 5 (1975).
39. H. Stroppe, *Wiss. Z. TH Magdeburg* **15**, 473 (1971).
40. S. Hähle, D. Pötzschke, H. Beyrich, and H. Markas, *Krist. Techn.* **11**, 91 (1976).
41. R. A. Keeley and M. T. Sprackling, *J. Phys. D: Appl. Phys.* **9**, 615 (1976).
42. B. H. Kear and P. L. Pratt, *Acta Met.* **6**, 457 (1958).
43. Ch. Zaminer, *Ber. Dtsch. Keram. Ges.* **42**, 73 (1965).
44. L. M. Davies, *Proc. Br. Ceram. Soc.* **6**, 1 (1966).
45. See, e.g., L. Föppl and E. Mönch, *Praktische Spannungsoptik*, Springer-Verlag, Berlin and New York (1972).
46. J. Kvapil, *Krist. Techn.* **4**, 123 (1969).
47. J. Kvapil, J. Sulovsky, and J. Kvapil, *Krist. Techn.* **7**, 1169 (1972).
48. K. H. Brauer, J. Feuerstake, F. Fröhlich, and U. Mohr, *Krist. Techn.* **8**, 253 (1973).
49. Y. J. Sirotin, *Sov. Phys.–Crystallogr.* **1**, 556 (1956).
50. K. A. Parsons, *J. Appl. Phys.* **24**, 469 (1953).
51. R. C. O'Rourke and A. W. Saenz, *Qt. Appl. Math.* **8**, 303 (1950).
52. C. N. Reid, *Deformation Geometry for Materials Scientists*, Pergamon, Oxford (1973).
53. J. B. Wachtman and L. H. Maxwell, *J. Am. Ceram. Soc.* **37**, 291 (1954).
54. R. Scheuplein and P. Gibbs, *J. Am. Ceram. Soc.* **45**, 439 (1962).
55. H. Palmour III, *Proc. Br. Ceram. Soc.* **6**, 209 (1966).
56. K. C. Radford and C. W. A. Newey, *Proc. Br. Ceram. Soc.* **9**, 131 (1967).
57. J. Grabmaier and C. Watson, *Phys. Status Solidi* **23**, K7 (1968).
58. L. Merker, *J. Am. Ceram. Soc.* **45**, 366 (1962).
59. A. H. Heuer and J. P. Roberts, *Proc. Br. Ceram. Soc.* **6**, 17 (1966).
60. R. Falckenberg, J. Electrochem. Soc. **123**, 63 (1976).
61. D. A. Curtis and J. S. Thorp, *Br. J. Appl. Phys.* **16**, 734 (1965).
62. J. S. Thorp, D. A. Curtis, and D. R. Mason, *Br. J. Appl. Phys.* **15**, 775 (1964).
63. K. Shiroki, *Jap. J. Appl. Phys.* **6**, 121 (1967).
64. W. Seifert, Thesis, University of Munich (1969).
65. S. M. Wiederhorn, B. J. Hockey, and D. E. Roberts, *Phil. Mag.* **64**, 783 (1973).
66. J. B. Wachtman and L. H. Maxwell, *J. Am. Ceram. Soc.* **42**, 432 (1959).
67. H. S. Bagdasarov, G. V. Berezhkova, V. G. Govorkov, E. P. Kozlovskaya, E. A. Fedorov, and M. A. Chernysheva, *Krist. Techn.* 8, 507 (1973).
68. W. Seifert, *J. Crystal Growth* **12**, 17 (1972).
69. K. Shiroki, *Rev. Sci. Instrum.* **38**, 1541 (1967).
70. R. S. Mitchell, *Rev. Sci. Instrum.* **36**, 1667 (1965).
71. R. H. Arlett and M. Robbins, *J. Am. Ceram. Soc.* **50**, 273 (1967).
72. L. M. Belyaev, C. Barta, A. A. Popova, A. F. Zakatov, and J. Zemlicka, IVth Conference on Single Crystals, Turnov, CSSR (1961).
73. Fabricated by Metals Research, Melbourne, England.
74. Unpublished work at Institute of Crystallography, Moscow; see Chapter 4 by C. H. L. Goodman following this chapter.
75. R. C. Pastor and A. C. Pastor, *Materials Res. Bull.* **1**, 275 (1966).
76. A. C. Pastor and R. C. Pastor, *Materials Res. Bull.* **2**, 555 (1967).

77. A. Goldsmith, T. E. Waterman, and H. Y. Hirschhorn, *Handbook of Thermophysical Properties of Solids*, Vol. 3, Macmillan, New York (1961).
78. J. F. Lynch, C. G. Ruderer, and W. H. Duckworth, Engineering Properties of Ceramics, Defense Documentation Center, Alexandria, Virginia.
79. E. Ryshkewitch, *Oxide Ceramics*, Academic Press, New York (1960).
80. C. Barta, J. Zemlicka, and V. Kment, IVth Conference on Single Crystals, Turnov, CSSR (1961).
81. A. G. Jones, *Analytical Chemistry*, Butterworth, London (1959).
82. D'Ans-Lax, *Taschenbuch für Chemiker und Physiker*, Springer-Verlag, Berlin (1970).
83. W. H. Bauer and W. G. Field, in *The Art and Science of Growing Crystals*, J. J. Gilman (ed.), Wiley, London (1963).
84. D. Haberland, *Mber. Dtsch. Akad. Wiss. Berlin* **4**, 376 (1962).
85. M. Yamamoto, S. Arino, and T. Sato, *Sci. Rep. Res. Inst. Tohoku Univ.* **22**, 156 (1971).
86. N. Peters, *Int. J. Heat Mass Transfer* **19**, 385 (1976).
87. P. A. Libby and C. Economos, *Int. J. Heat Mass Transfer* **6**, 113 (1963).
88. B. Lewis and G. von Elbe, *Combustion Flames and Explosions of Gases*, Academic Press, New York (1971).
89. K. Annamalai and P. Durbetaki, *Combustion and Flame* **25**, 137 (1975).
90. H. Rörtgen, Thesis, Aachen Univ. (1971).
91. N. I. Ikornikova and A. A. Popova, *Dokl. Akad. Nauk SSSR* **106**, 460 (1956).
92. J. A. Adamski, *J. Appl. Phys.* **36**, 1784 (1965).
93. R. A. Lefever and G. W. Clark, *Rev. Sci. Instrum.* **33**, 769 (1962).
94. Ch. H. Moore, *Trans. Am. Inst. Mining Met. Eng.* **184**, 194 (1949).
95. J. Ricard, Produits Chimiques Usine Kuhlmann—Département Rubis Synthètique des Alpes, Jarrie/Isère, France.
96. N. A. Velikhova, *Kristallografiya* **8**, 804 (1963).
97. K.-Th. Wilke, *Kristallzüchtung*, VEB Deutscher Verlag der Wissenschaften, Berlin (1973).
98. O. Lauer, *Feinheitsmessungen an technischem Stäuben*, Alpine AG, Augsburg (1963).
99. See, e.g., Physik der Adhäsion, *Int. Colloq. Chem. Ing. Technol.* **41**, 1276 (1969).
100. H. Rumpf, *Pharm. Ind.* **34**, 270 (1972).
101. R. Müller, H. Quart, and R. Warm, *Krist. Techn.* **5**, 589 (1970).
102. R. K. Verma, G. N. Sirkar, and S. Chatterjee, *Sci. Ind. Res.* **13A**, 516 (1954).
103. M. Kestigian, *Rev. Sci. Instrum.* **33**, 1293 (1962).
104. R. E. Carter, *Rev. Sci. Instrum.* **34**, 588 (1963).
105. J. A. Adamski, *Rev. Sci. Instrum.* **40**, 1634 (1969).
106. J. L. Sampson, *Rev. Sci. Instrum.* **42**, 278 (1971).
107. R. Falckenberg, *Krist. Techn.*, to be published.
108. J. L. Bazhenova, E. B. Zeligman, and S. N. Shorin, *Izv. Akad. Nauk SSSR, Neorg. Mater.* **8**, 869 (1972).
109. A. A. Popova, *Rost. Kristallov.* **4**, 168 (1965).
110. O. N. Boksa, S. V. Grum-Grzymailo, A. A. Popova, and E. F. Smirnova, *Kristallografiya* **13**, 725 (1968).
111. A. A. Popova and W. B. Sotkina, *Dokl. Akad. Nauk SSSR* **169**, 92 (1966).
112. A. A. Popova, *Sov. Phys.—Crystallogr.* **2**, 711 (1958).
113. R. C. Pastor, H. Kimura, L. Podoksik, and M. A. Pearson, *J. Chem. Phys.* **43**, 3948 (1965).
114. R. C. Pastor, A. C. Pastor, H. Kimura, and K. Avita, *J. Chem. Phys.* **44**, 4486 (1966).
115. F. K. Volynec and N. A. Cvetkova, *Neorg. Mater.* **6**, 271 (1970).
116. V. G. Sil'nichenko and M. M. Gritsenko, *Sov. Phys.—Crystallogr.* **9**, 647 (1965).

117. R. H. Hoskins and B. H. Soffer, *Phys. Rev.* **133A**, 490 (1964).
118. A. Kelley and R. B. Nicholson, eds., Strengthening Methods in Crystals, Wiley, New York (1971).
119. R. Müller, *Krist. Techn.* **5**, K29 (1970).
120. J. Kvapil, J. Sulocky, and J. Kvapil, *Krist. Techn.* **7**, 1169 (1972).
121. D. V. Sandreyev, P. A. Arsenyev, Z. G. Mareyeva, A. A. Mayer, R. M. Tolchinsky, and V. L. Farschtendiker, *Krist. Techn.* **8**, 957 (1973).
122. M. L. Kronberg, *Science* **122**, 599 (1955).
123. M. L. Kronberg, *Acta Met.* **5**, 507 (1957).
124. J. G. Grabmaier and R. Falckenberg, *J. Am. Ceram. Soc.* **52**, 648 (1969).
125. G. K. Bansal and A. H. Heuer, *Fracture Mechanics of Ceramics*, Vol. 2, R. C. Bradt, D. P. Hasselman, and F. F. Lange (eds.), Plenum Press, New York (1974).
126. R. Falckenberg, *Materials Res. Bull.* **8**, 171 (1973).
127. R. W. Cahn, *Physical Metallurgy*, North-Holland Publ., Amsterdam (1965).
128. A. Smakula, *Einkristalle,* Springer-Verlag, Berlin-Göttingen-Heidelberg-München (1962).
129. F. W. Harrison, *Res. Appl. Ind.* **12**, 395 (1959).
130. J. C. Brice, *The Growth of Crystals from Liquids*, North-Holland Publ., Amsterdam (1973).
131. A. Neuhaus and W. Richartz, *Ber. DKG* **35**, 108 (1958).
132. W. H. Bauer, J. Gordon, and C. H. Moore, *J. Am. Ceram. Soc.* **33**, 140 (1950).
133. W. H. Bauer, I. Gordon, and C. H. Moore, *Am. Mineralogist* **35**, 128 (1950).
134. C. Barta and R. Barta, *Zh. Prikl. Khim. Leningrad* **29**, 341 (1956).
135. A. M. Lejus and J.-P. Connan, *Rev. Int. Htes Temp. Refract.* **11**, 215 (1974).
136. A. M. Lejus, J.-C. Bernier, and R. Collongues, *Rev. Int. Htes Temp. Refract.* **11**, 183 (1974).
137. W. H. Bauer and J. Gordon, *J. Am. Ceram. Soc.* **34**, 250 (1951).
138. I. A. Bondar', A. A. Popova, M. M. Piriotko, and N. A. Toropov, *Dokl. Akad. Nauk SSSR* **175**, 1051 (1967).
139. H. Saalfeld, *Z. Kristallogr.* **133**, 396 (1971).
140. H. Saalfeld, *Am. Mineralogist* **60**, 824 (1975).
141. A. A. Popova, *Growth of Crystals*, Vol. 4, A. V. Shubnikov and N. N. Sheftal (eds.), Consultants Bureau, New York (1966), p. 120.
142. W. S. Brower and E. N. Farabaugh, *J. Appl. Phys.* **36**, 1489 (1965).
143. R. A. Lefever and J. Matsko, *Materials Res. Bull.* **10**, 281 (1975).
144. R. A. Lefever and G. W. Clark, *Rev. Sci. Instrum.* **33**, 769 (1962).
145. R. Dittman and D. Petzelt, *J. Crystal Growth* **23**, 77 (1974).
146. R. A. Lefever, *Rev. Sci. Instrum.* **33**, 1470 (1962).
147. A. A. Popova, *Sov. Phys.—Crystallogr.* **15**, 1060 (1971).
148. M. Kestigian, *Nature* **197**, 1006 (1963).
149. K. A. Wickersheim and R. A. Lefever, *J. Opt. Soc. Am.* **50**, 831 (1960).
150. R. H. Arlett, *J. Am. Ceram. Soc.* **45**, 523 (1962).
151. W. Fr. Eppler, *Z. Angew. Mineral.* **4**, 345 (1943).
152. K. Nakano, H. Tabata, H. Okuda, and N. Kogyo, *Gijutsu Shikensho Hokoku* **17**, 197 (1968).
153. C. C. Wang, *J. Appl. Phys.* **40**, 3433 (1969).
154. M. Yamamoto, S. Avino, and T. Sato, *Sci. Rep. Res. Inst. Tohoku Univ.* **22**, 156 (1971).
155. D. Haberland, *Mber. Dtsch. Akad. Wiss. Berlin* **4**, 376 (1962).
156. F. A. Reiss, *Appl. Optics* **5**, 1902 (1966).
157. J. Kvapil, *Krist. Techn.* **5**, 551 (1970)

158. J. Grabmaier and C. Zaminer, *Z. Angew. Phys.* **17**, 26 (1964).
159. J. Ricard and A. Cioccolani, *J. Crystal Growth* **13/14**, 718 (1972).
160. C. Barta, *Rost Kristallov* **6**, 181 (1965).
161. G. W. Dueker, C. M. Kellington, M. Katsmann, and J. G. Atwood, *Appl. Optics* **4**, 110 (1965).
162. S. Sh. Gendelev, *Sov. Phys.—Crystallogr.* **8**, 731 (1964).
163. J. Kvapil and J. Sulovsky, *Krist. Techn.* **6**, 769 (1971).
164. J. Kvapil, B. Pernev, J. Kvapil, and J. Sulovsky, *Czech. J. Phys. B* **24**, 389 (1974).
165. E. R. Dobrovinskaya and L. A. Litvinov, *Krist. Techn.* **6**, K33 (1971).
166. M. O. Kliya, *J. Crystal Growth* **3/4**, 719 (1968).
167. D. L. Stephens and W. J. Alford, *J. Am. Ceram. Soc.* **47**, 81 (1964).
168. H. Palmour III, J. J. DuPlessis, and K. Wurth Kriegel, *J. Am. Ceram. Soc.* **44**, 400 (1961).
169. J. Grabmaier and Chr. Watson, *Z. Angew. Phys.* **24**, 108 (1968).
170. C. Barta, F. Petru, and B. Hajek, *Naturwiss.* **45**, 36 (1958).
171. R. H. Gilette, *Rev. Sci. Instrum.* **21**, 294 (1950).
172. E. J. Scott, *J. Chem. Phys.* **23**, 2459 (1955).
173. A. A. Popova, *Sov. Phys.—Crystallogr.* **2**, 711 (1958).
174. T. Nakamichi, T. Sato, and Y. Nagayama, *J. Japan. Inst. Met.* **40**, 366 (1976).
175. J. G. Bednorz and H. J. Scheel, *J. Crystal Growth,* **41**, 5 (1977).
176. P. Y. Kikin, *Kristallografiya* **20**, 673 (1975).
177. K. Dräger and H.-J. Studt, *J. Crystal Growth* **37**, 151 (1977).

Chapter 4

Subsidiary Electrical Heating for Verneuil Furnaces in the USSR

C. H. L. GOODMAN

A visit to the USSR in November 1974 enabled the author to see some of the remarkable work in Verneuil furnace design at the Institute of Crystallography of the Academy of Sciences in Moscow (Leninskii Prospekt 59) that appears to be almost completely unknown outside the Soviet Union. This Institute, apart from its interest in the structure of crystals, is also one of the major centers of research on crystal growth in the Soviet Union. Its work on the growth of alumina crystals, particularly ruby, under Professor K. S. Bagdasarov has a particularly high reputation. The Special Construction Bureau at the Institute has designed and put into small-scale production a massive Verneuil furnace with a rugged subsidiary electrical heater which seems to offer a number of advantages, particularly where the growth of large crystals is concerned. In view of Dr. Falckenberg's comments in Chapter 3 on the difficulties associated with the incorporation of subsidiary electrical heating in a Verneuil furnace, this apparently highly successful design seemed to merit special comment—hence this brief note.

The subsidiary heater in these furnaces is made up from thick tungsten rod about 4 or 5 mm in diameter. This is welded to give a cage-structure heater of some 10 cm bore and 60 cm long (see Fig. 1 for construction details). This is eventually placed outside a high-quality, gas-tight alumina ceramic furnace tube of special high purity. The entire assembly ends up inside what appears to be a stainless steel outer case some 50 cm in diameter, and presumably containing refractory, through which hydrogen gas is circulated to protect the heater against oxidation

C. H. L. GOODMAN · Standard Telecommunication Laboratories Ltd., London Road, Harlow, Essex, England.

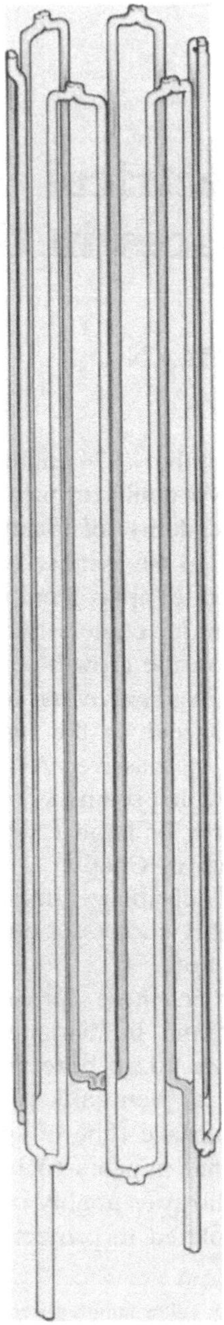

FIGURE 1. Cage-structure heater.

when in use. (It should be noted that this hydrogen obviously has to contain sufficient water vapor content to prevent the breakdown of the alumina ceramic furnace tube on account of the formation of volatile aluminum suboxide, Al_2O. It must not, however, be at too high a concentration lest it lead to massive formation of volatile WO_3 from the heater; i.e., it is probably present at about the 100–200 ppm level.)

The heater, of course, requires a low-voltage, high-current supply, which is furnished by a 35-kV A transformer mounted immediately below the furnace. The output voltage from this transformer is stabilized via a thyristor network to ±0.02% in order to obtain adequately stable temperatures within the furnace. Although no precise data were available regarding temperature stability, this was believed to be better than ±2°C at operating temperature.

The main Verneuil furnace room at the Institute has about 10 of these furnaces in operation, producing essentially cylindrical ruby crystals of up to at least 50-mm diameter and 400-mm length (i.e., considerably larger than those discussed by Dr. Falckenberg). Use of the subsidiary heater enables the maximum stress in such crystals to be kept in the range 2 to 4 kg/mm^2, which is claimed to be considerably below values typical for crystals grown in conventional Verneuil furnaces without subsidiary heating (see Chapter 3, Section 6.1.2.). It is also claimed that a further, postgrowth, high-temperature anneal followed by slow cooling in one of these furnaces, specially reserved for the purpose, and utilizing electrical heating alone, reduces the stress to a maximum of about 1 kg/mm^2. Such annealing is now standard practice, and, for example, gives crystals entirely suitable for high-power laser application. This obviously answers well for the quality and homogeneity of the material produced.

Acknowledgment

The author is indebted to Professor B. K. Vainshtein, Director of the Institute of Crystallography, who has approved the draft of the present text. It is earnestly hoped that an account of the fine work on crystal growth of ruby and sapphire at this Institute will be published in the literature in due course.

Index

GPSR Compliance

The European Union's (EU) General Product Safety Regulation (GPSR) is a set of rules that requires consumer products to be safe and our obligations to ensure this.

If you have any concerns about our products, you can contact us on ProductSafety@springernature.com

In case Publisher is established outside the EU, the EU authorized representative is:

Springer Nature Customer Service Center GmbH
Europaplatz 3
69115 Heidelberg, Germany

Batch number: 10682411

Printed by Printforce, the Netherlands